Manual of Water Supply Practices

M38

Electrodialysis and Electrodialysis Reversal

Second edition

American Water Works Association

Manual of Water Supply Practices—M38, Second Edition

Electrodialysis and Electrodialysis Reversal

Disclaimer

The authors, contributors, editors, and publisher do not assume responsibility for the validity of the content or any consequences of its use. In no event will AWWA be liable for direct, indirect, special, incidental, or consequential damages arising out of the use of information presented in this book. In particular, AWWA will not be responsible for any costs, including, but not limited to, those incurred as a result of lost revenue. In no event shall AWWA's liability exceed the amount paid for the purchase of this book.

If you find errors in this manual, please email books@awwa.org. Possible errata will be posted at www.awwa.org/M38.

Senior Managing Editor/Project Manager: Melissa Valentine
Technical Editor: Dianne Beirne
Cover Image: Dmitriy Rybin/Shutterstock.com
Publishing Operations Manager/Technical Illustrations: Gillian Wink
Production: Senior Specialist - Manuals: Willadee Hitchcock

Library of Congress Cataloging-in-Publication Data

Names: Wood, Jonathan H., author. | American Water Works Association, issuing body.
Title: Electrodialysis and electrodialysis reversal / by Jonathan H. Wood.
Other titles: AWWA manual ; M38.
Description: 2nd edition | Denver, CO : American Water Works Association, [2022] | Series: AWWA manual ; M38 | Includes bibliographical references and index. | Summary: "Directed to engineers and operators of ED and EDR systems, this manual provides detailed background information on ED and EDR as they relate to water treatment processes. The manual explains process principles, equipment information, electrodialysis technology, and system design. Information on water chemistry is included to enhance understanding of water processing"-- Provided by publisher.
Identifiers: LCCN 2021061088 | ISBN 9781647170691 (paperback) | ISBN 9781613006030 (adobe pdf)
Subjects: LCSH: Saline water conversion--Electrodialysis process. | Electrodialysis. | Handbooks and manuals.
Classification: LCC TD480 .W66 2022 | DDC 628.1/674--dc23/eng/20220125
LC record available at https://lccn.loc.gov/2021061088

Printed in the United States of America

ISBN 978-1-64717-069-1 e-ISBN-13 978-1-61300-603-0

https://doi.org/10.12999/AWWA.M38ed2

American Water Works Association

American Water Works Association
6666 West Quincy Avenue
Denver, CO 80235-3098
awwa.org

Contents

This page intentionally blank.

Figures

This page intentionally blank.

Tables

This page intentionally blank.

Preface

This second edition of AWWA Manual M38 provides detailed information on the classical electrodialysis (ED) and the electrodialysis reversal (EDR) processes and systems. ED and EDR systems employ electrochemical and membrane cell technologies to separate ionic materials in aqueous solutions. These systems have proven useful in food processing, medical applications, and other specialized industrial uses, with major applications being the production of drinking water or pure industrial process water from brackish, saline, or mineralized sources.

Directed to engineers and operators of ED and EDR systems, this manual provides detailed background information on ED and EDR as they relate to water treatment processes. The manual explains process principles, equipment information, electrodialysis technology, and system design. Information on water chemistry is included to enhance understanding of water processing.

It is hoped that this manual will also assist water process engineers and treatment plant decision makers in understanding the value of ED and EDR technology applied to their water treatment needs.

This page intentionally blank.

Acknowledgments

The AWWA Technical & Educational Council, the Water Quality & Technology Division, and the Membrane Processes Committee gratefully acknowledges the contributions made by those volunteers who drafted, edited, and provided the significant and critical commentary essential to updating M38. The Technical Review Board members dedicated many hours in the final stages of preparation of this second edition to ensure the overall technical quality, consistency, and accuracy of the manual.

Technical Review Board Members
Jonathan Wood, chair, Evoqua Water Technologies, Pittsburgh, Pa.
Younggy Kim, McMaster University, Hamilton, Ont., Canada
Desmond F. Lawler, University of Texas at Austin, Austin, Tex.
Judith Herschell Cole, Herschell Environmental, Pittsburgh, Pa.
Bernie Mack, Veolia Water Solutions & Technologies, Natick, Mass.
Corinne Bertoia, AWWA, Denver, Colo.

Contributors to the 2nd edition
W. Shane Walker, chair, University of Texas at El Paso, El Paso, Tex.
Rick Bond, Black & Veatch, Kansas City, Mo.
Val Frenkel, Greeley and Hansen, San Francisco, Calif.
Patrick Girvin, Suez Water Technologies & Solutions, San Diego, Calif.
Ali Sharbat, California State Polytechnic University, Pomona, Calif.
Jamie Vinsant, SUEZ Water Technologies & Solutions, Burlington, Mass.
Jonathan Wood, Evoqua Water Technologies, Lowell, Mass.

This page intentionally blank.

Chapter 1

Introduction

By W. Shane Walker

Overview of the Electrodialysis Process

Electrodialysis (ED) is an electrically driven membrane process used to demineralize or desalinate brackish and saline water. Brackish waters lie under approximately two-thirds of the United States, and some inland rivers, such as portions of the Rio Grande and the Colorado, are also brackish. Water is generally classified as brackish when mineral content ranges between that of fresh drinking water and that of seawater; brackish water contains more than 500 mg/L of total dissolved solids (TDS), and seawater contains more than 30,000 mg/L TDS. ED systems are capable of treating variable source water quality while producing a consistent finished water quality.

According to the Global Water Intelligence (GWI) DesalData website (incorporating the IDA Desalting Plants Inventory), the installed worldwide capacity of ED membrane treatment plants increased from 2 mgd (315 m^3/h) in 1955 to more than 200 mgd (31,545 m^3/h) in 1992 and more than 758 mgd (119,556 m^3/h) in 2019.

Reduction of Salinity (Total Dissolved Solids)

An overview of membrane process target applications based on size and molecular weights of contaminants is shown in Figure 1-1. ED processes remove total dissolved solids in the form of ions, or salinity, which may also be removed by reverse osmosis (RO) or nanofiltration (NF). RO operates by passing *water* through the membrane and rejecting dissolved contaminants, whereas ED processes operate by passing *ions* (charged compounds) through the membranes (as water flows parallel to the membranes). As a result, ED systems are generally more robust with respect to fouling and scaling than RO, and ED systems can generally be designed for higher recovery than RO systems.

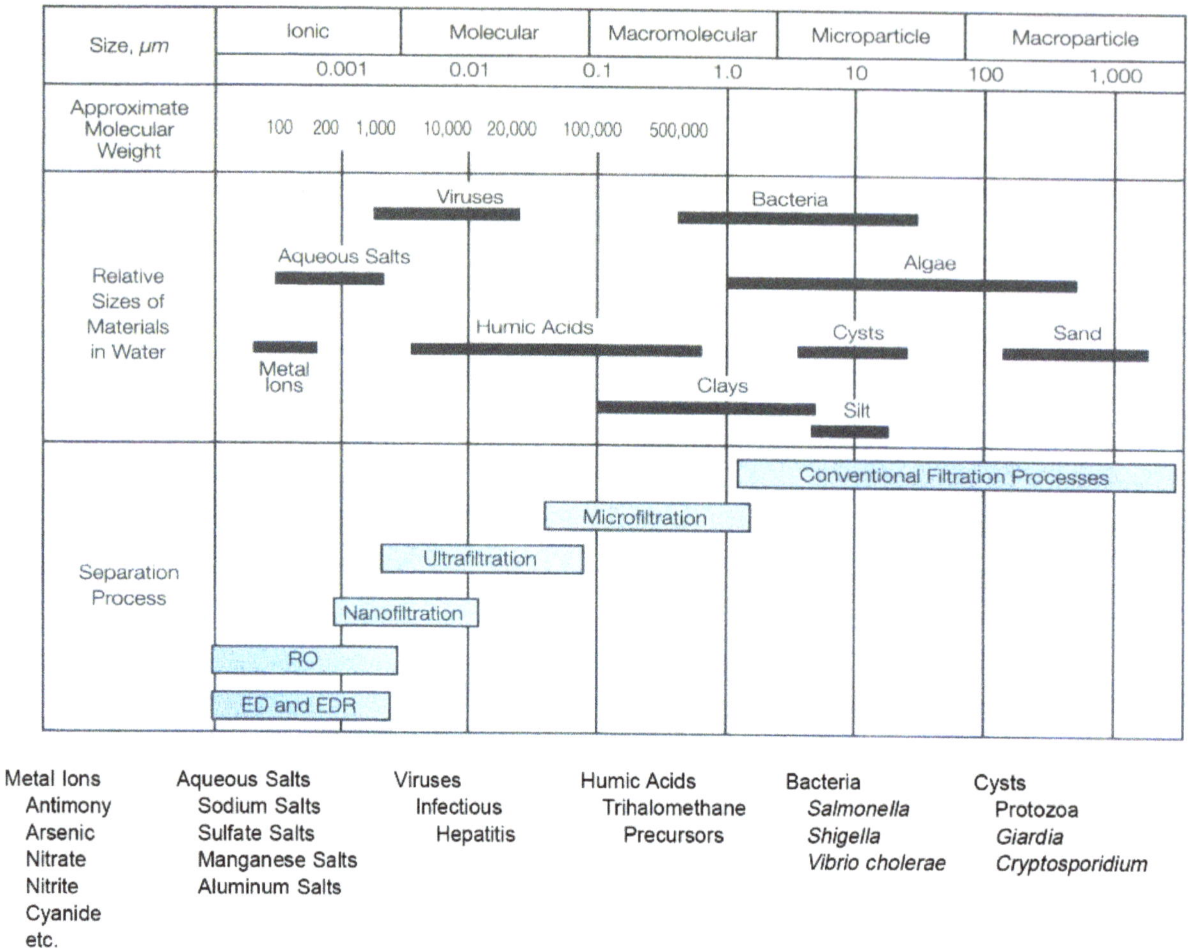

Figure 1-1 Membrane processes overview

While RO and NF processes typically remove a fixed fraction (e.g., 90–99 percent) of dissolved solids, electrodialysis plants can be designed to remove a targeted fraction (typically from 50–99 percent) of dissolved solids. Source water salinities of 100 mg/L to greater than 100,000 mg/L TDS can be successfully treated with ED with tailored salinity removal ratios of 50–90 percent, and ED can also be used to treat brackish concentrates, as well as seawater (Tanaka 2015). In addition to overall TDS reduction, ED and EDR systems can effectively treat source water with problems such as saltwater intrusion; high hardness; high nitrate/nitrite and selenium levels from agricultural contamination; high, naturally occurring arsenic or fluoride levels; and heavy metals contamination.

Review of Water Chemistry

A basic understanding of the composition of solutes in water whether salts, ions, or polar compounds is necessary to understand the design, operation, and maintenance of an electrodialysis treatment system. A brief review of water chemistry concepts is provided here, but several texts are available that provide more in-depth discussion of the topic (Bard et al. 2001, Newman & Thomas-Alyea 2004, Robinson & Stokes 2002, Wright 2007).

Table 1-1 Common ions found in natural waters

Cations (+)	Anions (-)
Sodium (Na^+)	Chloride (Cl^-)
Potassium (K^+)	Bicarbonate (HCO_3^-)
Magnesium (Mg^{+2})	Carbonate (CO_3^{2-})
Calcium (Ca^{+2})	Fluoride (F^-)
Strontium (Sr^{2+})	Sulfate (SO_4^{-2})
Barium (Ba^{2+})	Nitrate (NO_3^-), Nitrite (NO_2^-)
Iron (Fe^{2+}, Fe^{3+})	Phosphate (PO_4^{3-})

Ionic Solutions

An *ion* is a charged atom (e.g., sodium) or polyatomic compound (e.g., sulfate), and a mixture of ions in water is an electrolyte solution. An *electron* is a subatomic particle that, by convention, has an assigned charge of negative one (–1). A neutral atom that gains an electron will have a net charge of –1, and conversely, a neutral atom that loses an electron will have a net charge of +1. These charged atoms are called *ions*. Positively (+) charged ions are called *cations*, and negatively (–) charged ions are called *anions*. These two types of ions can dissociate in water. Some common ions that may be found in natural waters are shown in Table 1-1.

Ionic compounds, such as salts, minerals, acids, and bases (e.g., sodium chloride, NaCl, or table salt), are composed of charged atoms (e.g., Na^+ and Cl^-) that are held together by electrically attractive forces. If a crystal of sodium chloride is dissolved in water, the two atoms separate into two electrically charged particles, one with a positive charge (e.g., Na^+) and one with a negative charge (e.g., Cl^-). In addition to the ions shown in Table 1-1, ED may also be used to remove ionic species of metals such as cadmium, chromium, copper, mercury, selenium, and uranium.

Electrical Conductivity

With respect to ED, an important property of an ionic solution is its ability to conduct electricity. When two electrodes are connected to a direct current (DC) power supply (e.g., a battery) and immersed in pure water, negligible electric current passes between the electrodes because of a lack of ions in the solution to transport the electrical current. In a solution with sufficient concentration of ions, however, the dissociated ions are forced to move by the electric field, and their movement constitutes the electric current between the two electrodes. The ability of a solution to carry an electric current is known as electrical *conductivity*, and the conductivity of fresh and brackish water is typically reported in units of microsiemens per centimeter (μS/cm). The SI unit siemens is the reciprocal of an ohm, and the archaic term for this unit is the mho (ohm spelled backwards).

Conductivity is affected by the concentration of ions, the composition of ions, and the temperature of the solution in the following ways:

- Increasing ion concentration (salinity) increases electrical conductivity (i.e., decreases electrical resistance for concentrations less than several mol/L).
- Ions contribute differently to the overall conductivity depending on size and charge.
- Increasing the temperature increases ion mobility, resulting in an increase in conductivity.

Operating Principles of Electrodialysis

Electrodialysis (ED) is an electrochemical separation process in which ions are transferred through ion exchange membranes by means of an applied DC voltage. The ED process was developed in the 1950s and '60s (Mintz 1963, Solan et. al. 1971, Sonin & Probstein 1968) and has been described more generally over the past half-century (Benjamin & Lawler 2013, Davis et. al. 1997, Green & Perry 2018, Metcalf & Eddy 2014, Strathmann 2004, Tanaka 2015).

Ion Exchange Membranes

A basic electrodialysis stack is composed of two electrodes (cathode and anode) and repeating pairs of ion exchange membranes (cation and anion), as shown in Figure 1-2. The ion exchange (more correctly called *ion transfer*) membranes used in ED are described as either cation exchange (transfer) or anion exchange (transfer) membranes. These membranes are thin and flat with a relatively smooth surface, similar in appearance to a plastic sheet. The ions that are being removed diffuse through the water that is trapped inside the membranes, pulled by the electric field. The water that is being treated flows across the surface of the membranes; it is not pushed through the membranes as is the case in pressure-driven membrane processes.

Cation transfer membrane. The cation transfer membrane is a sheet composed of cation exchange media. This membrane allows the passage of cations while blocking the passage of water and anions, as shown in Figure 1-2. The membrane contains fixed, negative charges, which repel the negative ions while allowing the positive ions to pass through the membrane.

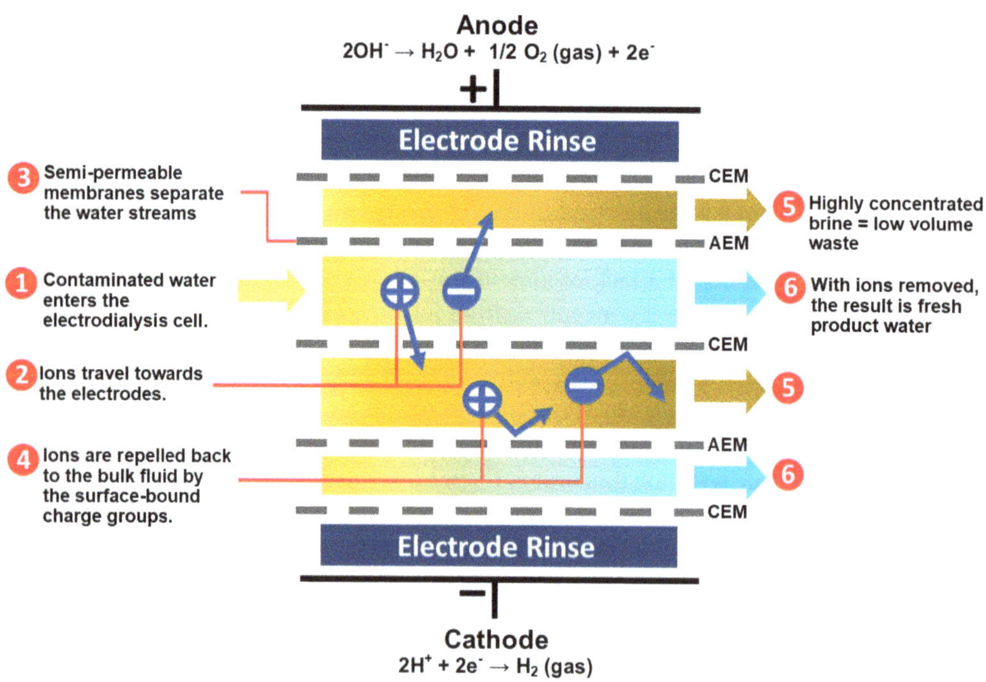

Adapted from Strathmann 2004

Figure 1-2 Electrodialysis stack operation (adapted from Strathmann)

Anion transfer membrane. The anion transfer membrane is a sheet composed of anion exchange media. This membrane allows the passage of anions while blocking passage of both cations and water. The membrane contains fixed, positive charges, which repel positively charged ions and allow negatively charged ions to pass through the membrane.

Operation

As feedwater flows parallel to the membranes (as shown in Figure 1-2), positively charged ions (e.g., Na^+ and Ca^{2+}) are attracted toward the negatively charged cathode, and negatively charged ions (e.g., Cl^- and SO_4^{2-}) are attracted to the positively charged anode. The desalination (separation) processes is described for each of the compartments:

- *Electrode rinse.* Cells 1 and 8 contain electrodes where reduction (cathode) and oxidation (anode) reactions occur.
- *Diluate.* Cells 2, 4, and 6 — Anions (e.g., Cl^-) are attracted toward the anode and pass through the anion membranes (A) into adjacent concentrate cells, while cations (e.g., Na^+) are attracted to the cathode and pass through the cation membranes (C) into adjacent concentrate cells.
- *Concentrate.* Cells 3, 5, and 7 — Anions (e.g., Cl^-) are attracted toward the anode but are repelled by the cation membranes (C) and are retained in the concentrate cells; while cations (e.g., Na^+) are attracted to the cathode but are repelled by the anion membranes (A) and are retained in the concentrate cells.

Thus, compartments 2, 4, and 6 are depleted of ions (known as *diluate* cells), whereas compartments 3, 5, and 7 have increased concentration of ions (known as *concentrate* cells). The flow through all of the diluate cells is collected in a diluate manifold to produce one effluent diluate (demineralized, product) stream, and the flow through all of the concentrate cells is collected in a concentrate manifold to produce one effluent concentrate (waste) stream. In full-scale operation, up to several hundred of these compartments may be assembled into a membrane stack.

Metal electrodes may be used to drive the electrical current through the stack by generating redox reactions at the electrode surface. At the cathode (-), water is reduced to consume electrons and produce hydrogen gas and hydroxide ions. At the anode (+), water is oxidized to produce electrons, oxygen gas, and hydrogen ions (protons); some chloride may be oxidized to chlorine gas (see Chapter 2). Because of the acidic and oxidizing environment of the anode compartments, anodes are normally constructed of inert materials, such as platinum-coated titanium. Some researchers are exploring the use of non-Faradaic (capacitive) electrodes for electrodialysis, but these are not yet widely used commercially.

Configurations of electrically driven membrane processes

Electrodialysis (Batch and Continuous)

The first type of commercial ED system was the batch system. In this ED system, source water is recirculated from a holding tank through the demineralizing spacers of a single membrane stack and back to the holding tank until the final purity is obtained. The production rate is dependent on the dissolved minerals concentration in the source water and on the degree of demineralization required. The concentrate stream is also recirculated to reduce wastewater volume, and continuous addition of acid might be required to prevent membrane stack scaling, depending on the ionic composition of the source water.

The second type of commercially available system was the continuous-flow ED. The demineralized stream makes a single pass through the stack and exits as product water. Higher salinity removal ratios can be achieved by arranging multiple stacks in series with the diluate effluent from the one stage entering the diluate to the following stage. The concentrate stream may be partially recycled to reduce wastewater volume, and scaling may be prevented by dosing the concentrate with antiscalant compounds and/or acid (depending on feed water composition). Control of scale and fouling is critical to all membrane systems — ED, EDR, RO, etc. Costs to install, operate, and maintain chemical feed systems as well as chemical storage facilities can significantly add to the costs of any membrane-based system.

Electrodialysis Reversal (EDR)

EDR systems, first developed in the 1960s, incorporate electrical polarity reversal to control membrane scaling and fouling. These systems are designed to produce demineralized water quasi-continuously without continuous chemical addition during normal operation.

In EDR systems, the polarity of the electrodes is typically reversed two to four times per hour. When polarity is reversed, the charge of the electrodes is reversed (i.e., the cathode switches to the anode, and the anode switches to the cathode). Thus, the diluate cells become concentrate cells, and the concentrate cells become diluate cells. After reversing polarity, valves in the system automatically switch so that water flowing through the new diluate cells now goes to the process, while water flowing through the new concentrate cells now goes to waste (or recycle).

In EDR, all electrodes must be tolerant of both the anode and cathode conditions. The polarity-reversal process affects the operation of a membrane system by

- detaching polarization films
- breaking up freshly precipitated scale or seeds of scale before they can cause damage
- reducing slime formations on membrane surfaces reducing problems associated with the use of chemicals
- cleaning electrodes with acid generated during anodic operation

Colloidal particles or slimes that are slightly electronegative may accumulate on the anion membrane and cause membrane fouling. This problem is common to most classes of ED systems. These fouling agents are removed by flushing with cleaning systems or by periodic polarity reversal as previously described.

Electrodialysis with permselective membranes

Permselective membranes can be tailored to inhibit the passage of divalent anions or cations, such as sulfates, calcium, and magnesium. Permselective refers to the ability of an ED membrane to discriminate between different ions allowing passage or permeation through the membrane. For example, some membranes show good permeation (high transport numbers) for monovalent anions, such as Cl– or NO3–, but have low transport numbers and show significantly lower permeation for divalent or trivalent ions, such as SO_4^{2-}, PO_4^{3-}, or similar anions, as shown in Figure 1-3. This is achieved by treating the anion membrane, and the effect can be exploited to separate various ions. Existing commercial membranes include monovalent anion specific, monovalent cation specific, or hydrogen ion specific. The relative specificities vary, with the monovalent anion membrane showing the greatest specificity (e.g., the ratio of chloride to sulfate ion transport numbers). Figure 1-3 shows the use of a standard membrane with a monovalent anion permselective membrane; the concentrate stream is very low in sulfate (i.e., approximately equal to or slightly greater than the feedwater), while the chloride and sodium concentrations, for example, could be many times higher than the feedwater.

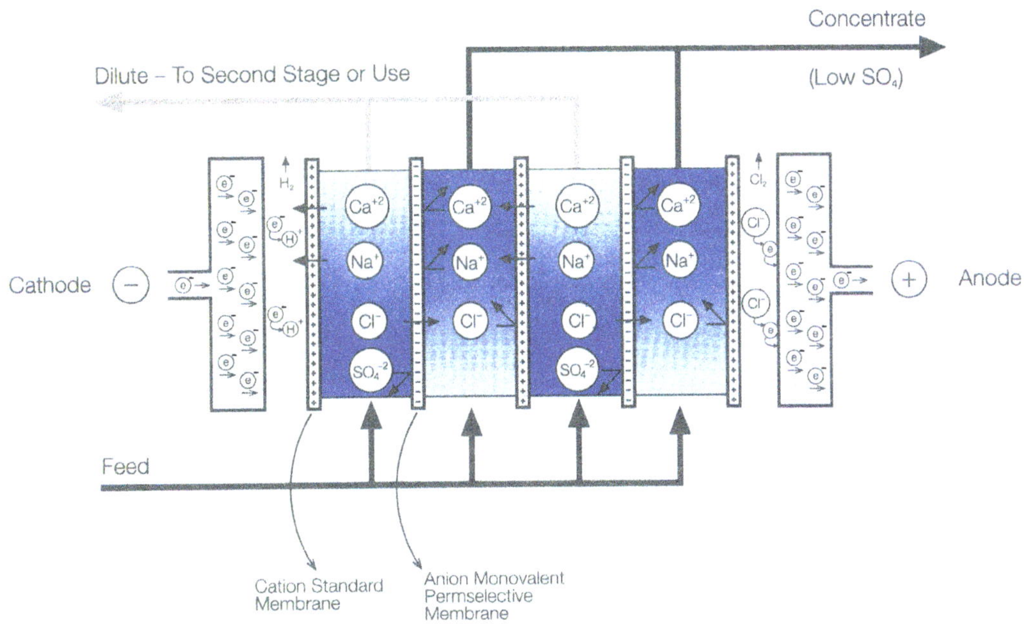

Source: Thomas D. Wolfe

Figure 1-3 Principle of monovalent permselective electrodialysis

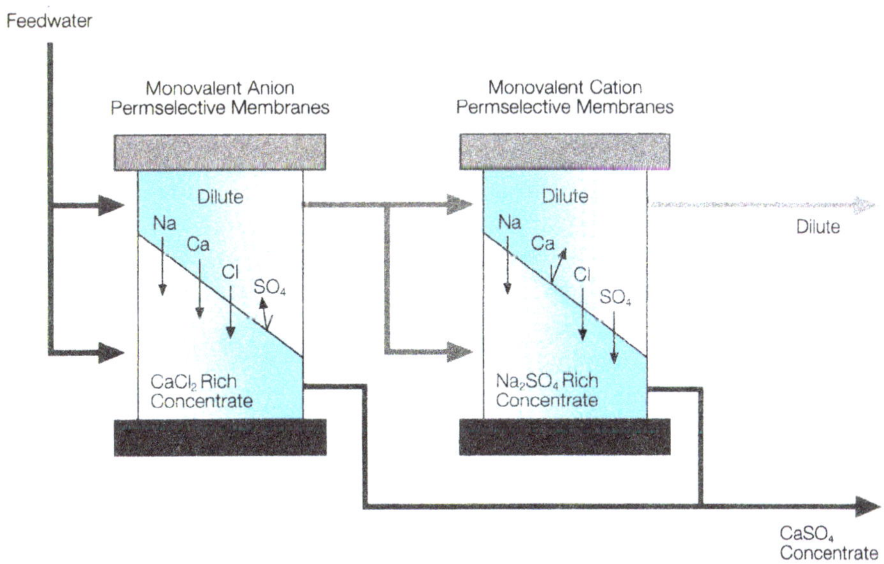

Source: Thomas D. Wolfe

Figure 1-4 Use of monovalent permselective ED membranes for high recovery (e.g., concentration of calcium sulfate in saturated waters)

Through the use of proper staging, with monovalent and divalent permselective membranes, the development of high calcium sulfate in the concentrate side of the membranes can be forestalled and scale formation prevented, as illustrated in Figure 1-4. In the first stage, relatively little sulfate passes through the membrane, and so the concentrate is rich in calcium chloride. Rather than passing this concentrate to stage 2, the stage 2 system concentrate is made up from fresh feed or another source. Here, the passage of calcium ions is retarded, and the concentrate is rich in sodium sulfate. Neither stage ever exceeds the calcium sulfate solubility limits. Yet, when the two concentrate streams are combined, together they can far exceed the calcium sulfate limit. In fact, precipitation can result on mixing.

In other schemes, the concentrate can be made up from a separate water source that is already low in sulfate (for example, reverse osmosis permeate or ED dilute water) to increase water recovery.

Electrodeionization (EDI)

Electrodeionization (EDI) was first commercialized in 1987 by the Process Water Division of Millipore Corporation (now Evoqua Water Technologies). EDI is a well-established process for the production of ultrapure water for several markets, including pharmaceutical/biotechnology, power generation (boiler feed water), and microelectronics/semiconductor. The process of EDI is a hybrid of two well-known separation processes—electrodialysis and ion exchange deionization. Like ED, EDI uses alternating anion-exchange and cation-exchange membranes, with spaces between the membranes configured to create liquid flow compartments with inlets and outlets. A direct current (DC) electric field is applied by an external power supply using electrodes at the ends of the membranes and compartments. The main construction difference between ED and EDI is that in EDI some or all of the compartments between the ion exchange membranes contain an electroactive, conductive filler—typically ion-exchange resin beads, as shown in Figure 1-5. For this reason, EDI has also been referred to as *filled-cell electrodialysis*.

The reason for the construction difference is a functional difference between the two processes. While ED is a bulk salt removal process typically used to produce drinking water from a brackish water feed, EDI is a polishing process most often used to make ultrapure deionized water from reverse osmosis permeate. In ED/EDR, when the feed solution is very dilute, the electrical resistance of the ED stack increases significantly, and it becomes difficult to pass DC and therefore, to remove the last traces of salt. In ED, the practical lower limit of product water conductivity is around 20 μS/cm—which is a typical feed water conductivity for an EDI system. The resin filler in EDI devices provides two modes of ion removal that allow the EDI process to produce effluent approaching the theoretical minimum water conductivity of 0.055 μS/cm at 25°C. In the first mode, transport of ions in the dilute solution is enhanced by the electroactive media, which helps conduct the ions to the surface of the membranes and then into the concentrating compartments (illustrated in Figure 1-5). The second mode involves operating with sufficient DC power to cause splitting of water into hydrogen and hydroxyl ions, thus electrochemically regenerating a portion of the ion exchange resin filler in the product water compartments. The water splitting and electroregeneration allow for the removal of weakly ionized species such as carbon dioxide, silica, and boron—which are not effectively removed by ED.

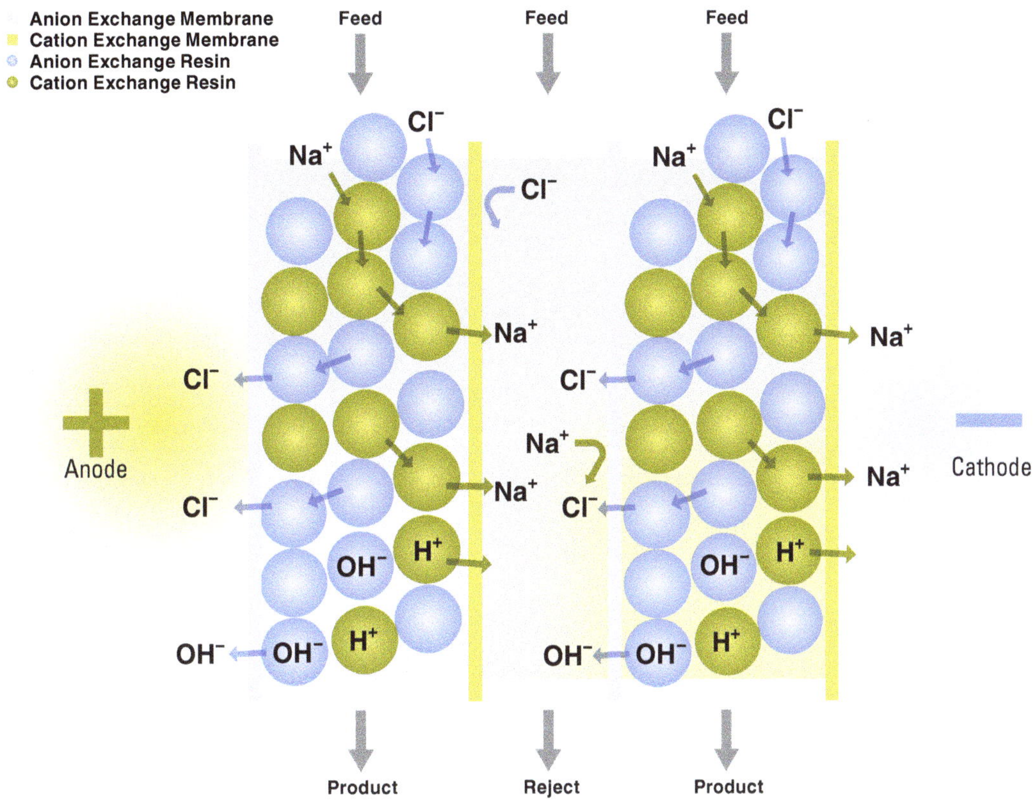

Figure 1-5 Ion transport and electrochemical regeneration in an EDI cell

REFERENCES

Bard, A.J.; Faulkner, L.R.; White, H.S. 2001. *Electrochemical methods: Fundamentals and Applications*. Wiley, New York.

Benjamin, M.M. and Lawler, D.F. 2013. *Water Quality Engineering*. Wiley, Hoboken, New Jersey.

Davis, T.A.; Genders, D.; Pletcher, D. 1997. *A First Course in Ion Permeable Membranes*. Electrochemical Consultancy, Hants, England.

Green, D. and Perry, R. 2018 (9th ed.). *Perry's Chemical Engineer's Handbook*. McGraw-Hill, New York, NY.

Metcalf & Eddy | AECOM. 2014. *Wastewater Engineering: Treatment and Resource Recovery*. McGraw-Hill, New York, NY.

Mintz, M.S. 1963. Electrodialysis—principles of process design. *Industrial and Engineering Chemistry*, 55(6), 18–28.

Newman, J.S. and Thomas-Alyea, K.E. 2004. *Electrochemical systems*. Wiley, Hoboken, New Jersey.

Robinson, R.A. and Stokes, R.H. 2002. *Electrolyte Solutions: Second Revised Edition*. Dover Publications, Inc. New York.

Solan, A.; Winograd, Y.; Katz, U. 1971. An analytical model for mass transfer in an electrodialysis cell with spacer of finite mesh. *Desalination*, 9(1), 89–95.

Sonin, A.A. and Probstein, R.F. 1968. A hydrodynamic theory of desalination by electrodialysis. *Desalination*, 5(3), 293–329.

Strathmann, H. 2004. *Ion-exchange membrane separation processes*. Membrane science and technology series, Elsevier, Amsterdam.
Tanaka, Y. 2015 (2nd ed.). *Ion exchange membranes: Fundamentals and applications*. Membrane science and technology series, Elsevier, Amsterdam.
Wright, M.R. 2007. *An Introduction to Aqueous Electrolyte Solutions*. Wiley, West Sussex, England.

Chapter 2

Applications and Process Design

By Jonathan Wood

Both ED and EDR are electrically driven membrane processes that selectively remove soluble ionic constituents carrying electrical charges that pass through semi-permeable ion exchange membranes. The natural electrical conductivity of water allows ED and EDR processes to be applied to a wide range of water treatment objectives.

In ED and EDR systems, the water being treated flows past, not through the membranes, so these systems do not present a barrier to remove bacteria or uncharged contaminants. In contrast, RO and NF systems pass water through membranes designed to reject contaminants in the molecular and ionic size ranges, effectively rejecting cysts (such as *Giardia* and *Cryptosporidium*), enteric viruses and bacteria. RO can be used in combination with ED and EDR to remove these contaminants and to further concentrate the waste stream.

Chapter 1 introduced conventional ion exchange membranes, also called ion-transfer membranes. These have a single type of charge (positive or negative) that allows them to exclude ions of the same charge. There are also membranes that contain both positive and negative charges and are thus called *bipolar membranes*. Applications for each are described in the following section.

APPLICATIONS OF MONOPOLAR MEMBRANE ELECTRODIALYSIS

Brackish Water Desalination

Reduction of TDS to meet drinking water standards is the most common application of ED and EDR technology. Plants treating brackish sources that contain up to 10,000 mg/L TDS can reliably and economically yield product water containing less than 500 mg/L TDS. For example, an EDR plant in Sarasota County, Fla., treats brackish well water with 2,500 mg/L TDS and yields product water of less than 350 mg/L TDS—an overall reduction in dissolved solids of 86 percent—with 85 percent recovery. The same plant can produce product water with 500 mg/L TDS from source waters containing as much as 3,600 mg/L TDS at the same rate of recovery. This flexibility is particularly important in applications for which multiple or variable source waters are used. The ability to control the product water quality by adjusting the applied DC voltage is an additional advantage of ED and EDR over other desalination processes such as reverse osmosis, which would typically employ blending to control product water quality to allow minimizing power consumption.

RO Reject Recovery

Electrodialysis can be used to treat the concentrate stream of reverse osmosis systems, in order to increase overall system water recovery. Because electrodialysis operates by removing ions from brackish water (instead of removing water, as in the case of reverse osmosis), in some cases higher recoveries can be achieved (Walker et al. 2014). This use of ED is of particular interest in zero-liquid-discharge systems, where it is desirable to reduce the volume of the RO brine stream due to the high cost of downstream brine concentrators. In some cases, the RO reject is softened before processing by electrodialysis, but polarity reversal has also been shown to be effective at preventing scaling of the ED (Wiesner et al. 2010). The fact that ED does not concentrate silica may be an advantage in some of these applications.

Seawater Desalination

While electrodialysis has been widely used for desalination of brackish water as previously mentioned, historically it has not often been applied commercially for the desalination of sea water. The most commonly stated reasons are high power consumption due to the high electrical resistance of the ion exchange membranes and diluting compartments as well as back diffusion due to the salinity difference between the concentrate and dilute steams.

In recent years, there have been technological advances that suggest there is a future for electrodialysis in desalination of seawater to make drinking water. Such improvements as counterflow, multi-stage operation, and the reduction of inter-membrane spacing was shown to reduce energy use to about 6 kW·h/m^3 (Turek 2002). With new lower-resistance ion exchange membranes, ED power consumption of 1.8 kW·h/m^3 has been demonstrated on a pilot scale (Shaw et al. 2011). This is comparable to the energy consumption of a state-of-the-art reverse osmosis system (Elimelech & Phillip 2011).

Table Salt Production

For many decades it has been common practice in Japan (which has no natural salt deposits) to use ED to concentrate seawater prior to evaporation to obtain salt crystals. The ED process can concentrate seawater from 3.5 percent by weight up to about 20 percent (near saturation). Acid is added to the seawater to minimize $CaCO_3$ scaling, and special monovalent selective membranes (which preferentially pass monovalent ions over polyvalents) help avoid $CaSO_4$ precipitation in the concentrate (Tanaka 2015).

Food Processing

Electrodialysis has found a number of applications in the food industry such as de-acidification of fruit juice, removal of tartaric acid from wine, and demineralization of products such as soy sauce, sugar molasses, amino acids, and cheese whey. By far, the largest market share of these applications is the demineralization of cheese whey and skimmed milk.

Whey is a byproduct that is produced in considerable amounts during production of cheese (about 9:1 whey to cheese ratio) and can present a waste disposal problem. While the whey is an excellent source of protein, lactose, vitamins and minerals, its high salt content makes it unsuitable as a food. ED has the ability to reduce the salt content of the whey without denaturing the proteins or causing decomposition of the vitamins. When the salt concentrations are reduced, the whey is similar in composition to human milk and useful in baby food products.

Desalting of whey or skimmed milk requires ion exchange membranes that are resistant to organic fouling. They must also be stable at both high and low pH, to tolerate the frequent chemical sanitization that is necessary to control growth of microorganisms. A typical system employs four electrodialysis stages (in series) and would achieve an overall salt reduction of 90 percent for whey and 70 percent for skimmed milk (Strathmann 2004).

Deionization

Electrodeionization (EDI) was first commercialized in 1987 by the Process Water Division of Millipore Corporation (now Evoqua Water Technologies). EDI (described in more detail in chapter 1) is now a well-established process for the production of ultrapure water for several markets, including pharmaceutical/biotechnology, power generation (boiler feed water), and microelectronics/semiconductor. EDI can treat water with relatively low electrical conductivity (e.g., 20 μS/cm) and produce high purity water approaching 0.055 μS/cm.

APPLICATIONS OF BIPOLAR MEMBRANE ELECTRODIALYSIS

Ion exchange membranes are selectively permeable. They contain fixed charges (anionic or cationic) and will allow the passage of counter-ions (opposite of the fixed charges) but not co-ions (of the same charge as the fixed charges). A cationic (positively charged) membrane will allow the passage of anions and is called an anion-permeable or anion-exchange membrane. An anionic (negatively charged) membrane will allow the passage of cations and is termed a cation-permeable or cation-exchange membrane. Normal electrodialysis stacks contain an alternating arrangement of separate anion exchange membranes (AEM) and cation exchange membranes (CEM).

A bipolar membrane contains an anion selective layer and a cation selective layer laminated together with some sort of transition region between the two. Application of an electric field will first remove salt ions from the transition region, and when these have been removed further transport of current can only be accomplished by hydrogen and hydroxyl ions resulting from splitting (dissociation) of water. The hydroxyl ions become available on the anode side of the bipolar membrane, and protons on the cathode side. Water is transported into the transition region from the solutions on either side of the bipolar membranes, as shown in Figure 2.1.

The process of bipolar membrane electrodialysis (BMED) combines bipolar membranes with monopolar anion-exchange and cation-exchange membranes to allow conversion of a salt to its corresponding acid and base. In such a case the repeating unit, or cell triplet, comprises a bipolar membrane, an acid compartment, an anion-exchange membrane, a salt compartment, a cation-exchange membrane and a base compartment. This arrangement is shown in Figure 2-2.

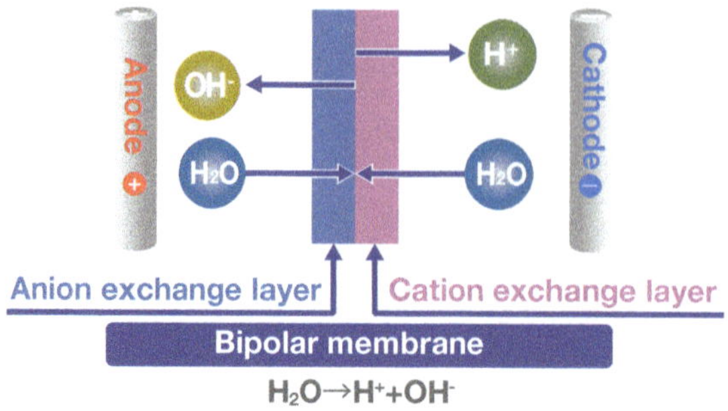

Source: ASTOM Corp. Website

Figure 2-1 Water splitting in a bipolar membrane

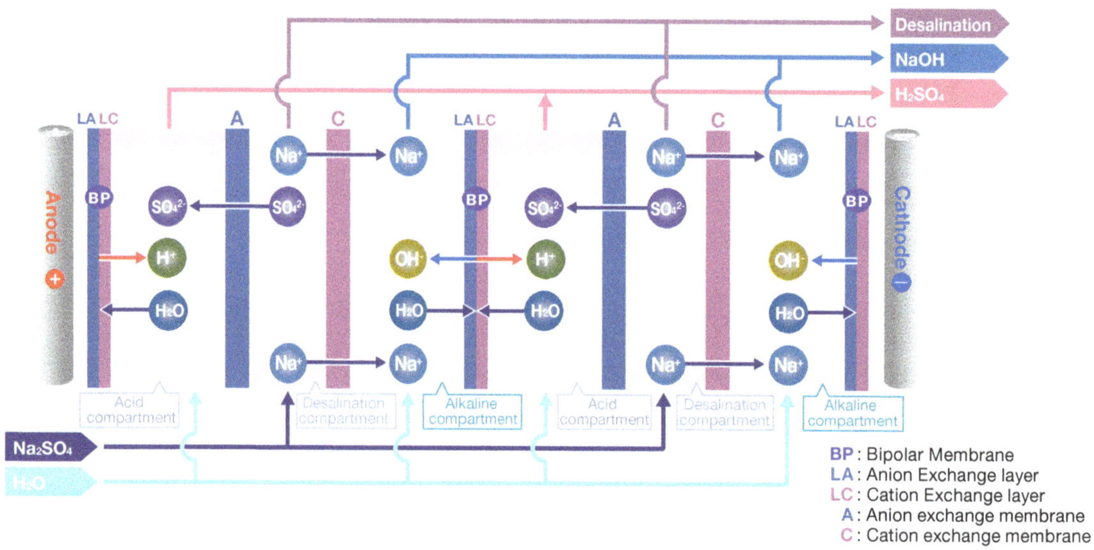

Source: ASTOM Corp. Website

Figure 2-2 Production of acid and base from salt by BMED

The use of bipolar membranes has been widely investigated, for such wide-ranging applications as production of acid and caustic for regeneration of ion exchange resins; recovery of organic acids from fermentation processes; removal of sulfur dioxide from flue gas; recycling of acids from steel pickling solutions; and dissociation of alcohols for production of alkoxides, to name a few. However, the BMED process has not been widely practiced commercially. The most frequently mentioned reasons are the relatively poor chemical stability of both the bipolar and monopolar membranes which results in fairly short service life, and poor permselectivity leading to product contamination (Strathmann 2004).

ALTERNATIVE SOURCES OF DC POWER

The most common way to power an electrodialysis stack is to take alternating current (AC) power from the grid through an electrical subsystem as discussed in chapter 4. The subsystem would consist of a rectifier to convert AC to the direct current (DC) power required by the electrodialysis process, as well as some means of controlling how much DC voltage and current is applied to the ED stack(s). However, many sources of renewable energy produce DC power, which can be efficiently used by the ED process.

Solar (Photovoltaic) Energy

The use of solar power to operate electrodialysis plants has been practiced in Japan since the early 1990s (Inoue & Kuroda 1993). Photovoltaic panels power the ED system directly through a control board. The photovoltaic array is designed with excess capacity, to allow operation of the ED system while simultaneously charging a storage battery system. The battery system is then used to power the ED during nighttime and in overcast weather.

Wind Energy

Unlike photovoltaics, wind turbines generally produce AC power that require the use of rectifiers to provide the DC power for electrodialysis. Like solar power, off-grid use of wind power presents challenges balancing the energy demand of the electrodialysis plant with the (variable) power produced by the windmills. The excess mechanical energy of the wind turbine can be stored using a flywheel to provide backup power (Veza et al. 2004). Flywheels have higher energy density than batteries and may be more environmentally friendly, but they are more complex and expensive and thus not as widely used. The National Renewable Energy Lab has tested a wind turbine to power an EDR system using battery energy storage (Corbus et al. 1999).

PRETREATMENT

ED/EDR may require some form of source water pretreatment for successful, long-term operation. As a minimum, suspended solids removal by 10-μm cartridge filtration is recommended. Additional pretreatment may be required, such as conventional coagulation, sedimentation, and filtration, depending on source water quality. ED and EDR normally require less pretreatment than NF or RO. Pretreatment methods for typical impurities encountered in source waters are summarized in Table 2-1.

Turbidity

The efficient operation of ED and EDR requires a source water turbidity level of no more than 2.0 ntu. When surface water turbidity exceeds 2.0 ntu, pretreatment may be required. Sand, multimedia, or other filtration may be necessary to achieve consistently low suspended solids or turbidity levels. Highly turbid sources may require in-line chemical coagulation and clarification followed by filtration.

Table 2-1 Typical water impurities and corresponding pretreatments

ED–EDR Impurity	Pretreatment	Tolerance Limit
Turbidity	Filtration	2.0 ntu
Iron	Oxidation and/or filtration	0.3 mg/L
Manganese	Oxidation and/or filtration	0.1 mg/L
Hydrogen sulfide	Oxidation	0.3 mg/L
Chlorine	Carbon adsorption or sodium bisulfite	0.5 mg/L as Cl_2 (continuous)

Source: Suez Water Technologies & Solutions

Most groundwater sources are consistently free of organic or suspended turbidity particles, making them suitable for direct ED and EDR treatment. The only pretreatment normally used with a groundwater source is an in-line 10-μm cartridge filter that is supplied as part of an ED or EDR system.

Iron and Manganese

Iron (Fe) and manganese (Mn) are most frequently found in groundwater from deep wells. Their concentrations may range from 0 to 15 mg/L. Iron and manganese may disrupt the ED process by fouling the membranes; maximum levels tolerated by ED and EDR are 0.3 mg/L Fe and 0.1 mg/L Mn. Pretreatment for iron and manganese includes oxidation by aeration or chemical addition followed by filtration; or a form of oxidizing filtration, such as manganese zeolite filtration or greensand.

Hydrogen Sulfide

A gas detectable by its characteristic "rotten egg" odor, hydrogen sulfide (H_2S) concentrations can range from 0 to 50 mg/L or higher in groundwater. The ED and EDR tolerance limit is 0.3 mg/L due to possible membrane fouling. There are several pretreatment processes that can be used for H_2S removal, including catalytic carbon or oxidation/filtration.

Chlorine

Chlorine may be present in some source waters as residual from upstream bacterial control. A significant advantage of electrodialysis over reverse osmosis is the tolerance to chlorine for limiting biofouling. Ion exchange membrane tolerance is 0.5 mg/L for free chlorine on a continuous basis. Membranes can be disinfected with a short-term dosage of 50 mg/L chlorine when necessary for bacterial control. Overall chlorine tolerance varies with membrane material and manufacturer. Typical pretreatment for chlorine removal is with carbon adsorption or sodium bisulfite injection.

Scale

The membrane recovery rate, or the percentage of process source water that is recovered as treated product water, can be limited by the presence of certain salts in the concentrate stream that can form a scale on the membrane surface. As water passes through the ED or EDR process, source water salts are removed and concentrated in the brine stream. The portion of concentrate that is wasted determines the concentration of salts in the brine or concentrate stream and the rate of recovery. In other words, repeatedly recycling the brine stream before sending it to waste creates a high concentration of salts in the brine stream and increases the recovery percentage. High water recovery minimizes wastewater pretreatment, reduces pumping costs, and may decrease waste disposal costs. A high concentration of salts, however, increases scaling potential if the concentration is beyond solubility limits. Calcium carbonate and sulfate salts of calcium, barium, fluoride, and strontium are the most common scale formers.

Potential for scale formation or membrane fouling also differs with the type of ED/EDR used. ED/EDR plants have strict limitations on concentrate chemistry and may require pre-softened source water or antiscalant addition to limit formation of deposits such as $CaCO_3$, $CaSO_4$, and $BaSO_4$.

ED systems can also employ special membranes to remove specific ions such as monovalent membranes to remove sodium chloride or nitrate. ED systems can also use two hydraulic stages, with anion monovalent permselective membranes in the first stage and cation monovalent permselective membranes in the second stage. By maintaining the concentrate streams from the two stages separate (and thus keeping the calcium and sulfate ions separate within the ED unit), the system can often reach effective calcium sulfate levels that are many times more than the normal saturation point when it finally mixes the two concentrate streams outside of the ED unit itself. Levels two to six times greater than calcium sulfate saturation can often be achieved, depending on the specific chemistry. Overall water recovery rates for such systems can reach as high as 95 percent on brackish waters.

The polarity-reversing EDR system requires less pretreatment than ED and is capable of higher water recovery. EDR's regular and automatic reversal of the DC polarity drives membrane foulants and scale deposits from membranes into the concentrate stream and out of the system. Without chemical antiscalants, the EDR concentrate stream can handle a Langelier Saturation Index (LSI) as high as +1.8 and a $CaSO_4$ saturation of 150 percent. Recovery rates can be further enhanced with antiscalant addition to the concentrate stream to achieve an LSI of +2.5 and a $CaSO_4$ saturation of 275 percent. This permits typical EDR systems to achieve water recovery rates of greater than 90 percent.

ED and EDR membrane systems can be cleaned in place to restore system performance after scale formation or reversible membrane fouling. This is discussed in detail in chapter 5.

POST-TREATMENT

Post-treatment requirements for ED and EDR systems depend on the final use of the product water (i.e., industrial or potable). Each application has different post-treatment needs.

Industrial users may require water with lower salinity than drinking water. Mixed-bed ion exchange units or electrodeionization (EDI) can remove more of the remaining ions from the ED and EDR product water, serving as *polishers* in high-purity applications.

In municipal or potable water applications, product water must meet both primary and secondary regulations established by the Safe Drinking Water Act. Thus, the post-treatment for ED/EDR product includes disinfection and corrosion control.

Potable ED and EDR plants are usually equipped with a small chlorinator to provide the level of disinfection required by federal and local health agencies. Chlorine is used most frequently to reduce pathogenic microorganisms, with ozone used for disinfection in some bottling applications.

Because of the corrosive nature of demineralized water, a stabilizing agent may be added to prevent or minimize oxidation of metallic surfaces. Commonly, the post-treatment objective is to maintain the LSI near zero or slightly positive to prevent corrosion of pipes. This may be accomplished by adding alkalinity through lime, soda, or caustic. A phosphate additive may also be used to minimize corrosion in metal pipes.

In some ways, electrically driven membrane processes such as ED/EDR differ significantly from pressure-driven membrane processes such as NF/RO, and the post-treatment requirements may vary as a result. Because ED/EDR can tolerate much higher concentrations of chlorine than RO, disinfection can be performed upstream, helping to avoid microbiological fouling of the ED/EDR. But because the water flows through the membrane in a pressure-driven process, these tend to be better for removal of particulates, organics, and chlorine-resistant pathogens such as viruses and cysts. ED/EDR also does not remove silica as effectively as RO, which can be either a disadvantage or a benefit (avoidance of scaling) depending on application.

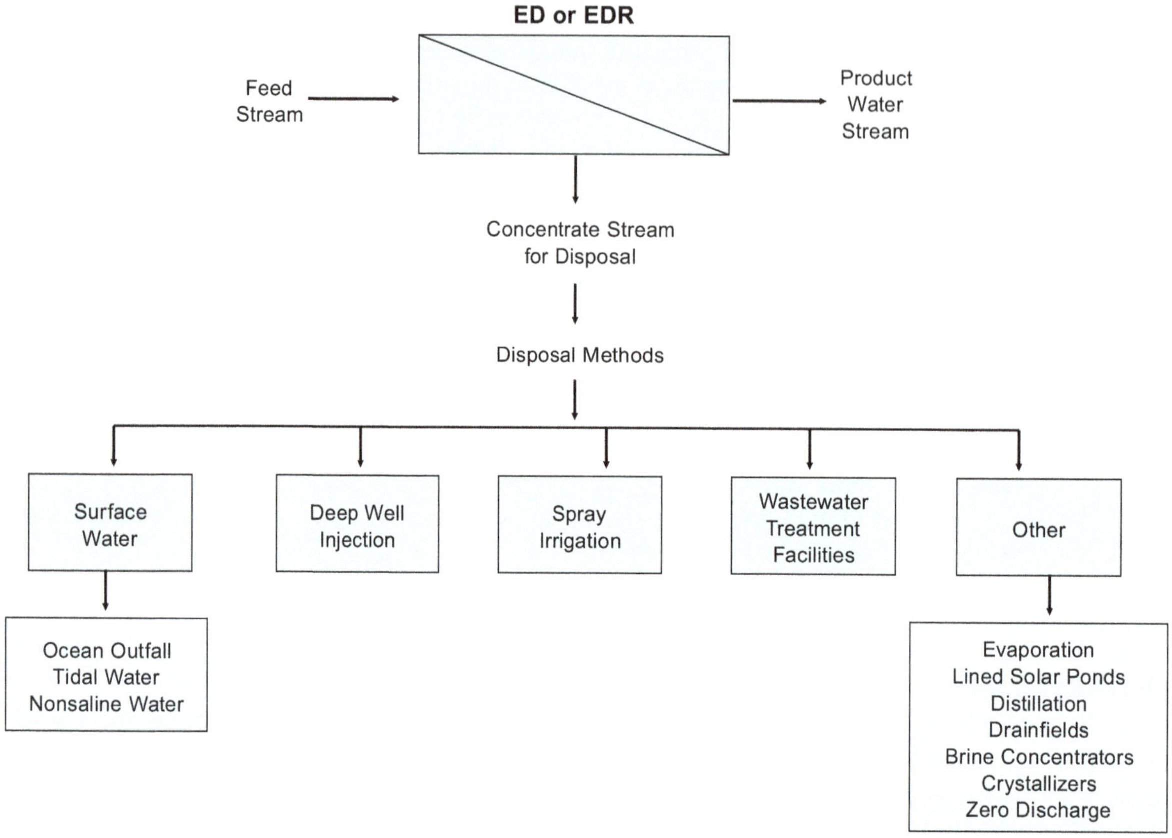

Adapted from Andrews (1991); Mickley (2006); and Mickley (2009)

Figure 2-3 Concentrate disposal methods

ED CONCENTRATE DISPOSAL

Concentrate is defined as the brine or waste stream from the membrane process. The method selected for concentrate disposal influences the initial cost of the system as well as operational expenses. This section discusses concentrate disposal methods, as illustrated in Figure 2-3 (Andrews 1991; Mickley 2006). It is important to note that processes such as ED and EDR with high recovery rates are associated with increased brine concentration.

Overview

Concentrate disposal permitting has become a major consideration for membrane process application as regulatory constraints for disposal of process wastes have increased. Table 2-2 describes the current required permits for four methods of disposal discussed in

Table 2-2 Examples of concentrate disposal permit requirements

Disposal Method	Required Permit
Surface water	National Pollutant Discharge Elimination System (NPDES) federal or state level permit; state or local discharge-to-surface water permit
Deep well injection	Underground Injection Control Program (UIC), regulated by EPA.
Spray irrigation	Federal, state, or local groundwater disposal permit
Wastewater treatment facility	Municipal or local treatment permits

this section—surface water, deep well injection, spray irrigation, and wastewater treatment facilities.

Pilot studies to predict concentrate quality determine when concentrate treatment is necessary for disposal. Concentrate quality may also be determined by means of a diffusion model if operating pressure, source water quality, recovery rate, and solvent and solute mass transfer coefficients are known (Conlon 1988).

Another factor influencing the selection of a disposal method is the concentrate flow rate Qc, which is determined from the following equation:

$$Q_c = (1 - r) \times Q_f$$

Where:

r = product water recovery (expressed as a decimal)

Qf = feed water flow rate

For example, consider a 1.2-mgd source water feeding to an EDR unit at 85 percent recovery (i.e., $Qf = 1.2$ mgd, $r = 0.85$). The concentrate flow rate may be calculated as follows:

$$\begin{aligned} Q_c &= (1 - r) \times Q_f \\ &= (1 - 0.85) \times 1.2 \\ &= 0.18 \text{ mgd} = 180{,}000 \text{ gpd} \end{aligned}$$

Methods of Concentrate Disposal

The following paragraphs describe current methods of concentrate disposal. Regulations vary for disposal, and the best option for one situation may not be appropriate for another. Concentrate disposal issues for ED/EDR are similar to those for NF/RO.

Surface water. Concentrate disposal to surface water includes transportation to ocean outfalls or tidal waters or to nonsaline waters. This method is considered the most cost-effective disposal approach. However, increasing regulations may make discharge to surface waters less feasible.

Concentrate discharge must meet surface water quality standards (including the addition of a mixing or dilution zone), cannot be acutely or chronically toxic, and must undergo toxicity testing. The concentrate must also meet anti-degradation regulations. Problem constituents in concentrate disposal to surface waters include hydrogen sulfide, dissolved oxygen, chlorides, radium 226, radium 228, gross alpha radioactivity, fluoride, and metals. The pH and conductivity of the concentrate are other relevant concerns.

The cost of disposal to a surface water is affected by the concentrate flow rate, the need for pretreatment to meet regulatory standards, transportation, operational costs, and outfall structures. Diluting concentrate with surface water in a mixing zone may be all that is necessary for regulatory compliance.

Deep well injection. This method of disposal involves injecting the concentrate into a well that is hundreds to thousands of feet deep. The injection zone contains high salinity natural fluids (of similar quality to the concentrate) and must be contained within impermeable layers of rock or other material.

Permitting for deep well injection requires that the concentrate cannot migrate to drinking water sources (TDS less than 10,000 mg/L) and that concentrate quality is compatible with aquifer water. Concentrate disposal is also regulated by geologic conditions and the industrial well classification.

Deep well injection involves strict well design and construction, requiring a corrosion-resistant interior liner and surrounding packer. High levels of TDS in the concentrate are mildly corrosive to steel.

Cost of disposal to deep wells is influenced by concentrate volume, transportation, injection-zone depth, pumping pressure, storage or alternate method of disposal, regulatory monitoring, and operational expenses. Operational costs include continuous monitoring of flow and wellhead pressure. Monthly testing may also be required to check for leaks.

Detailed geophysical logs are required before well construction. Permitting also requires television surveys of the well and radioactive tracer surveys to ensure well integrity before startup.

Spray irrigation. This disposal method uses concentrate as an irrigation resource applied by sprinkler, surface, or drip systems. Major considerations include the need for monitoring wells and the need for an alternate method of disposal during periods of heavy rain. Runoff control depends on the application system; sprinklers are the most effective. Concentrate salinity, toxicity, and soil permeability determine the feasibility of this disposal method.

Regulations governing groundwater quality and protection of drinking water aquifers must be investigated to confirm the acceptability of this alternative. Problem constituents include metals and radionuclides, as well as TDS and chlorides at levels unacceptable to plants and animals. Dilution of the concentrate is often required.

Factors determining the cost of disposal by spray irrigation include the concentrate flow rate, transportation, land acquisition and preparation, distribution, pumping pressure, concentrate storage for runoff, drainage, operation methods, and volume of runoff. The amount of runoff depends on which type of spray irrigation is used and may involve drainage canals or a reservoir, pump, and return pipeline. Operational costs are higher than for surface water or well injection methods because spray irrigation disposal is more labor intensive.

Wastewater treatment facilities. Disposal of concentrate to a wastewater treatment plant can cause the facility to exceed its treatment capability. Requirements for disposal include effluent disposal permits or sludge disposal permits, as well as compliance with federal limits on quantities of industrial waste. Problem constituents include chlorides, metals, and TDS.

Other methods. Concentrate disposal technologies used less frequently include the following:

- evaporation
- lined solar ponds
- distillation
- drain fields
- brine concentrators
- crystallizers

REFERENCES

Andrews, L.S. 1991. Disposal Methods and Issues of Concentrate Discharges — An Overview. In Proc. of the AWWA Membrane Conference in Baltimore, Md. Denver, Colo.: AWWA.

Conlon, W.J. 1988. Disposal of Concentrate From Membrane Process Plants. In Proc. of the AWWA Membrane Conference in Orlando, Fla. Denver, Colo.: AWWA.

Corbus, D.; Baring-Gould, I.; Drouilhet, S.; Gervorgian, V.; Jimenez, T.; Newcomb, C.; Flowers, L. 1999. Small Wind Turbine Testing and Applications Development. NREL/CP-500-27067

Elimelech, M. and Phillip, W.A. 2011. The future of seawater desalination: energy, technology, and the environment. *Science*, 333(6043), 712–717.

Inoue, S. and Kuroda, O. 1993. Electrodialysis desalination system power by photovoltaic power generation. *Industrial Application of Ion Exchange Membranes, vol. 2, Research group of electrodialysis and membrane separation technology, Soc. Sea Water Sci., Japan*, 177–180.

Mickley, M. 2009. "Treatment of Concentrate." Desalination and Water Purification Research and Development Program Report No. 155. U.S. Bureau of Reclamation. Available online.

Mickley, M. 2006 (2nd ed.). "Membrane Concentrate Disposal: Practices and Regulation." Desalination and Water Purification Research and Development Program Report No. 123. U.S. Bureau of Reclamation. Available online.

Shaw, M.; LeTourneau, T.; Liang, L.; Ng, K.H. 2011. Low Energy Desalination: Siemens Solution to a Global Challenge, IDA World Congress – Perth Convention and Exhibition Centre, Australia, Sep 4–9.

Strathmann, H. 2004. *Ion Exchange Membrane Separation Processes*. Elsevier, Amsterdam, Netherlands.

Tanaka, Y. 2015 (2nd ed.). *Ion Exchange Membranes: Fundamentals and Applications*. Elsevier, Amsterdam, Netherlands.

Turek, M. 2002. Cost Effective Electrodialytic Seawater Desalination. *Desalination*, 153(2002), 371–376.

Veza, J.M.; Penate, B.; Castellano, F. 2004. Electrodialysis desalination designed for off-grid wind energy, *Desalination*, 160, 211–221.

Walker, W.S.; Kim, Y.; Lawler, D.F. 2014. Treatment of Inland Brackish Groundwater Reverse Osmosis Concentrate with Electrodialysis – Part III: sensitivity to solution composition and hydraulic recovery. *Desalination*, 347, 158–164.

Wiesner, A.; Juby, G.; Zacheis, A.; Morquecho, R.; Mulvihill, T. Electrodialysis Reversal Provides Effective Treatment Approach for Concentrating Reverse Osmosis Brinee, WEFTEC 2010.

This page intentionally blank.

Chapter 3

Design

By Ali Sharbat

Design of an electrodialysis (ED) or electrodialysis reversal (EDR) plant is based on three key factors: the volume of water to treat, the constituents and temperature of the feed water, and product water requirements. Production water capacity requirements also determine the sizes of the pumps, pipes, and valves, as well as the number of membrane stacks in line. The degree of demineralization required dictates the membrane stack configuration or staging. Depending on the source water quality and desired demineralization, product water may be blended with untreated water, thereby lowering capacity and costs.

COMPONENTS OF ED AND EDR DESIGN

Prepackaged skid-mounted ED and EDR systems are available for production requirements ranging from a few thousand gallons (several thousand liters) per day to several million gallons (several million liters) per day. Mobile, trailer-mounted configurations are also available for temporary and permanent installation or for larger-scale pilot testing. These skid-mounted package plants minimize capital expenditure and engineering costs and simplify installation and training requirements.

Pilot studies may be required before new systems are implemented. These studies confirm computer model values, assist in developing specific design criteria, and refine cost data to allow the best process to be selected for a full-scale project.

Membrane Stacks

Most ED and EDR systems are designed specifically for a particular application. The amount of salt to be removed is determined by the configuration of the membrane stack. This section describes physical components of the membrane stack, which is the fundamental working unit of an ED or EDR system. A membrane stack may be oriented in either a horizontal or vertical position. Figure 3-1 shows a vertical stack, and a horizontal stack is shown in Figure 3-2.

Spaces between membranes represent flow paths formed by 0.02-in. to 0.04-in. (0.5-mm to 1.0-mm) thick, low-density polyethylene spacers. The dimensions of large spacers range from 18 in. × 40 in. (0.5 m × 1 m) to 3.3 ft × 6.5 ft (1 m × 2 m). Spacers may carry either demineralized or concentrated water because they are arranged in the stack so that the demineralized streams and the concentrate streams flow into separate single manifolds. In a typical system, each stack removes from 30 to 60 percent total dissolved solids (TDS) at flow rates of 100 to 130 gpm (6.3 to 8.2 L/s) and pressures of 50 psig (345 kPa).

Cell pairs form the basic building blocks of an ED membrane stack. A cell pair consists of the following:

- anion transfer membrane
- concentrate spacer
- cation transfer membrane
- dilute stream spacer

Source: SUEZ Water Technologies & Solutions and MEGA, respectively

Figure 3-1 Typical vertical SUEZ membrane stack

Figure 3-2 Horizontal MEGA stack

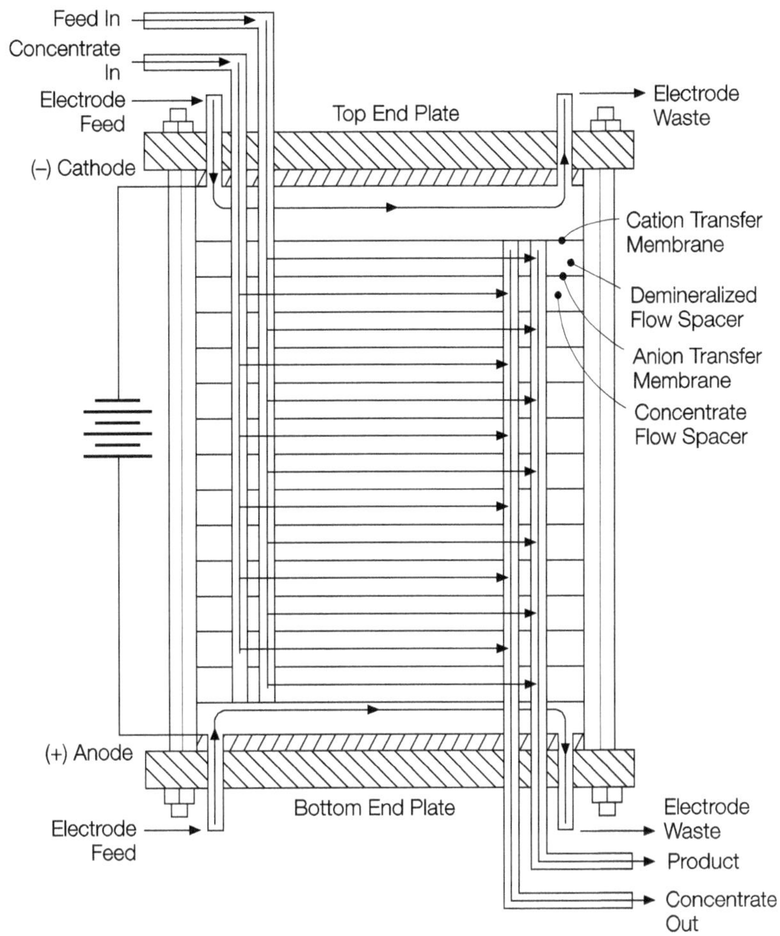

Source: SUEZ Water Technologies & Solutions

Figure 3-3 Membrane-stack flow paths

The number of cell pairs necessary to achieve a given product water quality is primarily determined by source water quality. A typical plant with an output of 50,000 gpd (189 m^3/d) is equipped with stacks containing 450 to 500 cell pairs.

Figure 3-3 illustrates the flow paths of the two types of streams through the membrane stack. The flows of these paths are as follows:

1. Source water (feed) flows parallel only through demineralizing compartments, whereas the concentrate stream flows parallel only through concentrating compartments.
2. As feedwater flows along the membranes, ions are electrically transferred through membranes from the demineralized stream to the concentrate stream.
3. Flows from the two electrode compartments do not mix with other streams. A degasifier vents reaction gases from the electrode waste stream, as explained later in this chapter.
4. Top and bottom plates are steel blocks that compress the membranes and spacers to prevent leakage inside the stack.

Membranes

The ion exchange membranes used in ED are described as either anion transfer or cation transfer membranes. These membranes are flat with a smooth surface, similar in appearance to a plastic sheet. Impermeable to water under pressure, membranes are usually reinforced with synthetic fiber. Both anion and cation transfer membranes have the following characteristics:

- low electrical resistance
- insolubility in aqueous solutions
- semirigid construction for ease of handling
- resistance to change in pH from 2 to 10
- ability to function continuously at ambient temperatures up to 115°F (46°C)
- resistance to osmotic swelling (volume expansion due to water uptake) in solutions between 220 mg/L and 60,000 mg/L salt
- long life expectancy (average 10 years for anion and 15 years for cation membranes)
- resistance to fouling
- impermeability to water under pressure
- availability in many sizes and configurations

The membranes are typically 0.001-in. to 0.024-in. (0.03-mm to 0.6-mm) thick and can be either homogeneous or heterogeneous.

Heterogeneous membranes are manufactured by mixing an ion-exchanging resin with a plastic binder. The ion exchanger should constitute at least half the mixture. This mixture is then heated and then either poured into a supporting open weave cloth or extruded through a die as either a supported or unsupported sheet.

Homogeneous membranes consist of a continuous ion exchange medium created by a copolymerization reaction that takes place in the presence of the supporting material. This eliminates the inert binder that is required for heterogeneous membranes.

With both homogeneous and heterogeneous membranes, dissolved ions can transfer through the membrane in either direction, depending on the type of ion and the orientation of the electric field.

Cation transfer membrane. This membrane allows the passage of cations while blocking the passage of water and anions. Negatively charged sites are incorporated into the membrane structure, and these repel the negative ions while allowing the positive ions to pass through the membrane. Figure 3-4 shows the chemical structure of a typical cation transfer membrane. As feedwater travels along the membrane surface, positively charged sodium ions (mobile counter cations) are attracted toward the negatively charged fixed anion sites of the membrane. Sodium ions are transported through the water-filled passage of the polymer-supported membrane and into the concentrate stream.

Anion membrane. The anion transfer membrane is similar in structure to the cation transfer membrane except that the charged sites incorporated into the membrane are cationic (see Figure 3-5). This membrane allows passage of anions like chloride while blocking passage of both cations and water.

Spacers

Spacers are typically die-cut polypropylene flow channels with manifold cutouts similar to those for membranes (see Figure 3-6). Manifold cutouts form pipes in the stack allowing water to flow between membranes through polyethylene netting/mesh (gray area in Figure 3-6). Spacers are alternately positioned between membranes in the stack to create independent flow paths.

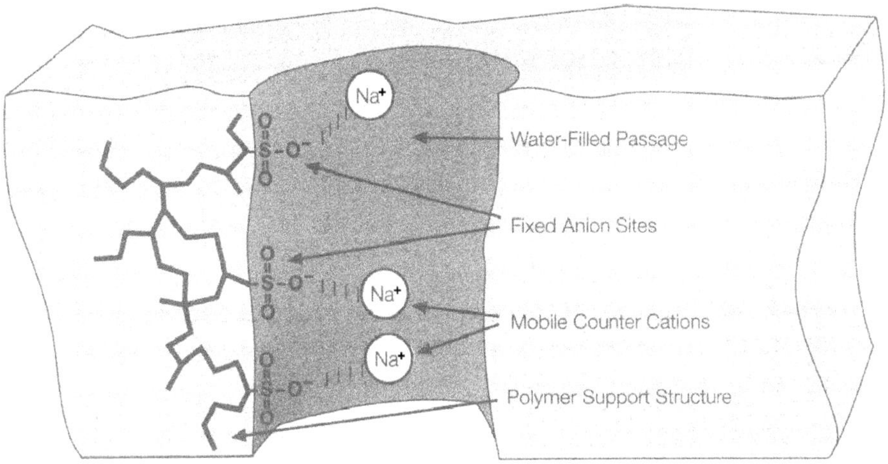

Source: SUEZ Water Technologies & Solutions

Figure 3-4 Structure of a cation membrane

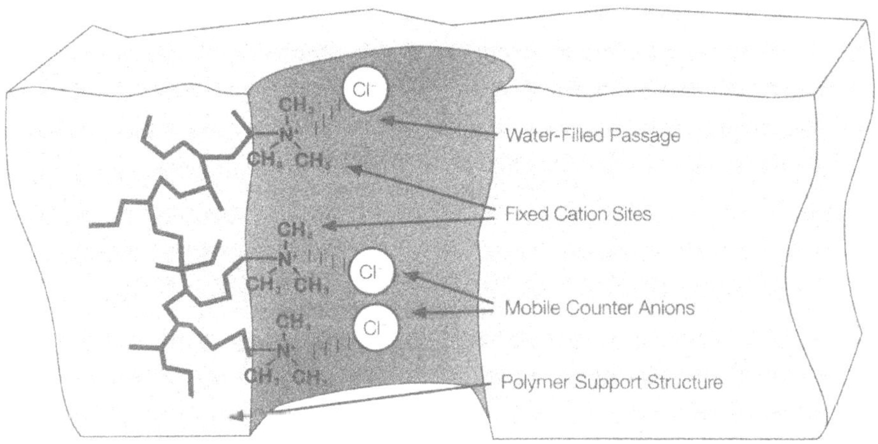

Source: SUEZ Water Technologies & Solutions

Figure 3-5 Structure of an anion membrane

Source: SUEZ Water Technologies & Solutions

Figure 3-6 Spacers

Diluate and concentrate spacers are created by rotating an identical spacer 180°. Diluate spacers allow water to flow across membrane surfaces where ions are removed, whereas concentrate spacers prevent the concentrate stream from contaminating the demineralized stream.

Increasing turbulence in the flow through the spacer promotes mixing of the water and the transfer of ions. Turbulence resulting from spacers also breaks up particles or slime on the membrane surface and attracts ions to the membrane surface. Flow velocities range from 0.1 to 0.3 ft/s (3 to 10 cm/s) for mesh spacers and from 0.5 to 1.6 ft/s (15 to 50 cm/s) for tortuous path spacers (Strathmann 2004). Higher velocities minimize concentration polarization, which increases the limiting current density, but higher velocities also increase the pressure drop through the stack. Maximum pressure for ED and EDR systems is generally limited to 50 psi (345 kPa), and pressure is lost at each stage of the system. Because pressure must be maintained throughout the system, the impact of spacers on pressure is an important design consideration.

Spacers create flow paths within the membrane stack for demineralized and concentrated water. Different thickness and mesh spacing of spacers satisfy specific design applications. Because the same spacers are used for both demineralized and concentrated water in EDR systems, the flow rates of both these streams should be equalized to prevent high differential pressures across the membranes (Siwak 1993).

Spacer design and flow velocity are also important parameters in determining the maximum amount of desalting or the current density (discussed later) at which a given system can operate.

Electrodes and Electrode Compartments

An electrode at each end of the membrane stack conducts direct current (DC) electric current into the stack. The electrode is usually made of titanium and plated with platinum. Its life span is dependent on the ionic composition of the source water and the amperage applied to the electrode. Large amounts of chlorides in the source water and high amperages reduce electrode life. Polarity reversal (as in EDR) also results in significantly shorter electrode lifetimes than for nonreversing systems.

Normally located at each end of the membrane stack, electrode compartments consist of an electrode, an electrode water-flow spacer, and a heavy cation membrane. The electrode spacer is thicker than a normal spacer, which increases water velocity to prevent scaling. This spacer also prevents the electrode waste from entering the main flow paths of the stack.

Effluent from these compartments may contain oxygen, hydrogen, and chlorine gas. Concentrate from the electrode stream may be sent to a degasifier to remove and safely dispose of any reaction gases.

STAGING

ED and EDR systems are designed to provide the required level of demineralization by staging the membrane stacks in series. Staging is designed to provide sufficient membrane area and retention time within the stack to achieve the desired level of demineralization. There are two types of staging in ED—hydraulic and electrical.

Hydraulic Staging

Hydraulic staging refers to increasing the number of passes the water makes along a membrane stack in an ED or EDR system. Maximum salt removal for any hydraulic stage is 50 to 67 percent, depending on the manufacturer, with typical design values of 40 to

65 percent. Increased salt removal is accomplished by the addition of more hydraulic stages in series (see Figure 3-7).

In systems that require higher capacities, additional stages are created by the addition of more membrane stacks in parallel. In small systems, hydraulic staging can be accomplished in a single stack by using one or more thick interstage membranes, as shown in Figure 3-8.

Electrical Staging

Electrical staging refers to the number of electrodes in a membrane stack. Normally, there is only one electrode pair in a membrane stack, but a second independent electrode pair could be inserted to provide maximum salt removal while avoiding polarization and hydraulic pressure limitations (which are explained later in this section). Electrical staging

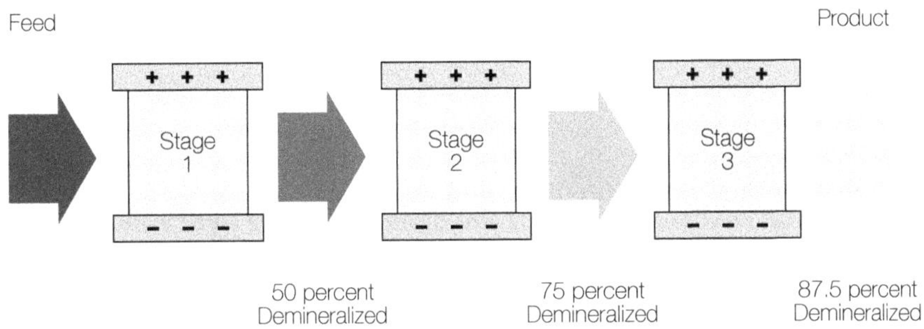

Source: SUEZ Water Technologies & Solutions

Figure 3-7 An ED system with three hydraulic and three electrical stages

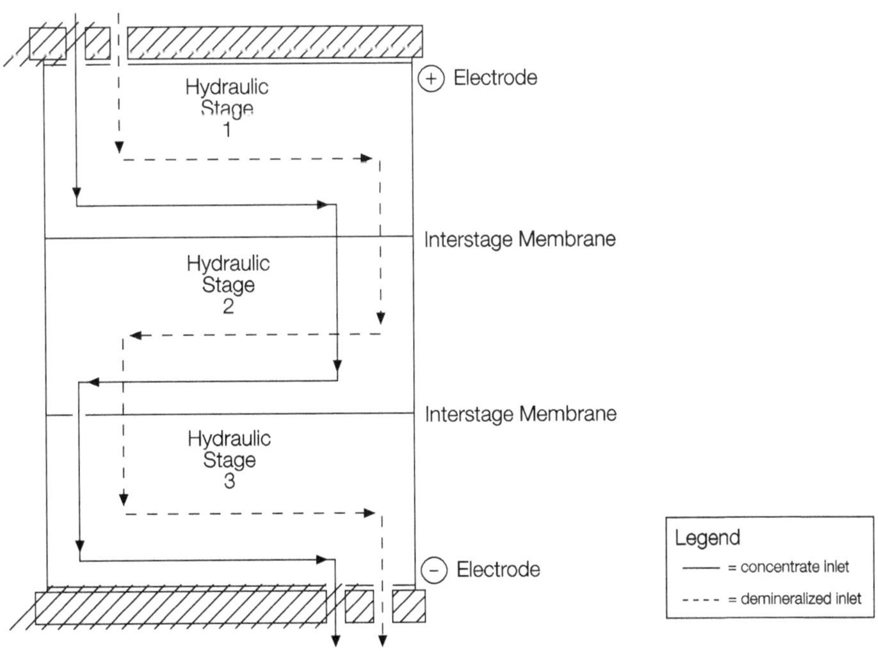

Source: SUEZ Water Technologies & Solutions

Figure 3-8 Hydraulic staging within a single membrane stack

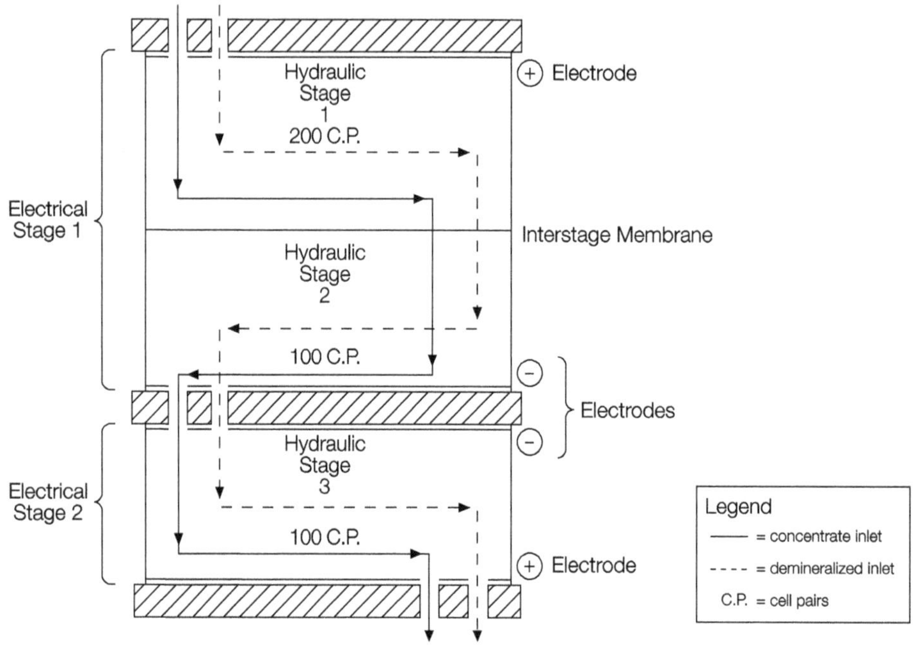

Source: SUEZ Water Technologies & Solutions

Figure 3-9 Membrane stack with two electrical stages and three hydraulic stages

applies independently controlled electrical current to each electrical stage to remove a certain amount of salt from the source water. Figure 3-9 illustrates an example of electrical staging.

LIMITING PARAMETERS

Several limiting parameters, which are discussed in the following paragraphs, need to be considered in the design of a successful plant.

Limiting Current Density (Polarization)

Current density can best be described as an electrical flux expressed in terms of amperes per square centimeter. It applies to the mobility of individual ions, and the limiting factor is the rate at which ions can freely move through a solution. Ions moving through the solution should balance ions moving through the membrane. Limiting current density is a function of fluid velocity in the flow path, stream temperature, and ionic composition.

Based on Boundary layer theory the movement of ions in the bulk of the solution passing between membranes is primarily influenced by turbulence and Brownian motion. Close to the surface of the membrane where velocities are lower, the motion of ions switches to being controlled by the applied current.

As current density increases, ions in a solution migrate through the membrane and boundary layer faster than they move from the bulk solution into the boundary layer. The boundary layer solution can become so depleted of ions that water molecules are affected. Increased voltage causes water to dissociate into hydrogen ions (H^+) and hydroxyl ions (OH^-). This is called the *polarization point* and occurs in the demineralizing compartments. When polarization becomes pronounced, the pH of the demineralized stream changes. Hydrogen ions and hydroxyl ions migrate through the cation and anion exchange membranes, respectively. In the concentrate stream, OH^- tends to combine with ions, such as calcium and magnesium. The resulting scale limits efficiency and requires frequent cleaning to prevent irreversible fouling in ED systems.

Operating at current densities sufficient to cause polarization results in the following for both ED and EDR systems:

- Energy consumption increases as electrical resistance increases.
- Current efficiency decreases because the transfer of hydrogen or hydroxyl ions is not the desired product.
- Additional increments of current become less efficient in transferring ions in polarization areas.

In commercial ED and EDR systems, a value of 70 percent of the limiting current density is used as the maximum allowable for brackish water operation, as illustrated in Figure 3-10. This provides a reasonable level of safety in system design.

High Salt Removal Rate EDR

For inland brackish water desalination using EDR, the two main areas of exploration that may end up to be big opportunities are reducing capital cost of EDR by reducing the size of EDR stacks and minimizing the volume of concentrate to reduce the cost of concentrate management. In both main areas, there are some niches in which process improvements would result in lower overall cost of desalination.

Sizing EDR equipment depends on several factors such as feed flow rate, feed chemistry, product requirements, and the flux of ions through ion exchange membranes in the EDR stack. Feed flow rate, feed chemistry, and product requirements are the factors dictated by the treatment plant, the flux of ions through membrane, however, can be modified by design engineers. Increasing the flux by increasing the applied energy (i.e., higher applied voltage and as a result higher current) corresponds to higher rate of salt removal. To reduce the capital cost of an EDR plant, salt removal rate (flux of ions through membrane surface) should be maximized. Understanding the limitations in increasing the

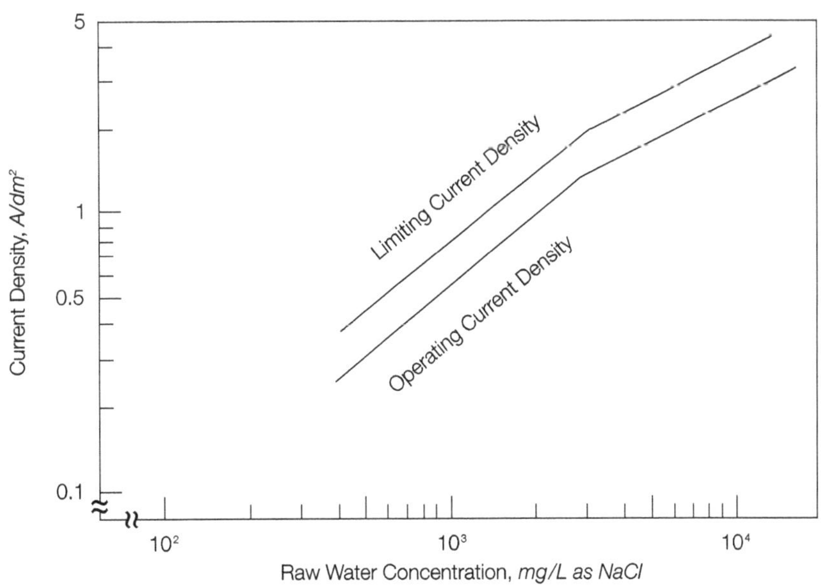

Source: SUEZ Water Technologies & Solutions

Figure 3-10 Raw water concentration and current density

flux helps the design engineers to reduce the membrane surface area and as a result reduce the capital cost of the EDR plant. As applied voltage to the EDR stack increases, current perpendicular to the membrane surface also increases and more ions are transferred from the boundary layer on the surface of the membrane in the dilute stream to the boundary layer on the surface of the membrane in the concentrates stream. However, by increasing the current density by increasing the voltage, the concentration of salt ions near the membrane surface in the dilute stream becomes depleted, and it becomes progressively more difficult to remove additional salt. Ultimately, at a certain current density, concentration of ions adjacent to the surface of the membrane in the dilute side becomes zero. This certain current density is called *limiting current density* and is one of the most important criteria in designing EDR systems. Reaching limiting current density has a number of adverse effects on salt removal: water dissociation into H^+ and OH^-, pH interruption, possible membrane damage, very high electrical resistance of the boundary layer and as a result high energy consumption. Different factors determine limiting current density in an EDR stack such as hydrodynamics of the flow (e.g., flow rate), temperature, feed water composition, and the shape and length of the spacer flow path.

Design engineers always try to maintain a safety margin to not get close to the limiting current density. However, the safety margin is not very well defined and the common safety margins used to design EDR systems might be too conservative. It appears that there is an opportunity that by increasing the conductivity reduction per stage, fewer stages are required to achieve a target overall salt removal, and this is a significant capital cost saving on the EDR desalination process.

Better understanding of limiting current density and the safety margin will assist design engineers to design EDR systems in a more efficient/aggressive manner, which would result in lower desalination cost. Recent research that has been performed on full-scale EDR unit at the Brackish Groundwater National Desalination Research Facility (BGNDRF) in Alamogordo, NM, demonstrates promising results that could potentially be used to increase the salt removal rate for each stage (Chastain 2011).

Current Leakage

Current leakage is defined as excess electric current at the electrodes that travels through an adjacent membrane, either an anion or cation membrane, into higher conductivity water, usually the concentrate. Figure 3-11 illustrates stack shorting or current leakage, with the stray current moving into the concentrate, then back toward the opposite electrode. The dotted line represents the normal movement of the DC current through the membranes and spacers without current leakage. Excess current can melt or "burn" the membranes and spacers. Normal design practices limit this voltage to 80 percent of the current that would cause burning. The limit is determined by water temperature, source water concentration, membrane stack size, and the internal manifold that splits flow into concentrate and dilute streams.

Back Diffusion

Back diffusion is the movement of ions from the concentrate stream to the demineralized stream against the force of the DC current. This occurs when the ion concentration in the concentrate stream is substantially higher than the ion concentration in the diluate stream. Methods for minimizing back diffusion include decreasing recovery and limiting polarization by lowering the applied DC current.

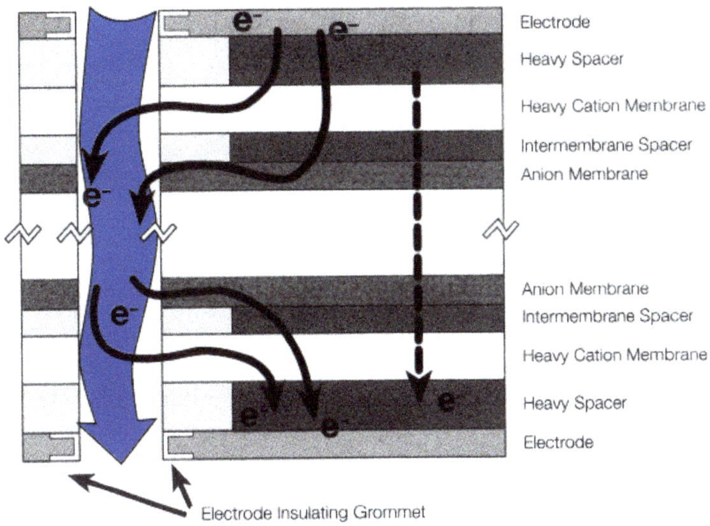

Source: SUEZ Water Technologies & Solutions

Figure 3-11 EDR stack shorting

Langelier Saturation Index

The Langelier Saturation index (LSI) is a measure of the scaling potential of a water source. Scaling is caused by the accumulation of calcium (Ca^{+2}), total dissolved solids, and bicarbonate (HCO_3^-) in the concentrate stream. The LSI is the difference between the actual pH and the pH at which scaling occurs. A positive LSI indicates a tendency for scaling, and a negative LSI indicates a nonscaling tendency (Pontius 1993). Design factors that may be adjusted to compensate for high LSI values include decreasing recovery, using an antiscalant, reducing pH by acid feed, or a combination of acid and antiscalant.

Calcium Sulfate Saturation

Control and prevention of calcium sulfate ($CaSO_4$) precipitation in an ED system are achieved by limiting the brine stream concentrations of calcium and sulfate. This is accomplished by decreasing the percent salt removal or percent water recovery, or by changing the ions removed. It can also be achieved by the addition of a sequestering agent or antiscalant agent, such as sodium hexametaphosphate, to the brine or concentrate stream. ED systems reach high water recovery by avoiding saturation conditions through the use of either scale inhibitors or permselective membranes (as discussed in chapter 1).

EDR systems, on the other hand, can satisfactorily handle calcium sulfate levels greater than saturation because of the polarity reversal, utilizing the calcium sulfate crystal induction time as the means of preventing calcium sulfate scaling. After polarity reversal the concentrate stream starts at near feed water quality and then gradually increases to the final concentrate concentration. The concentrate stays at the high point for only 15 to 30 minutes (less than the crystal induction time) before polarity reverses again, thus limiting the formation of scale. The upper design limits for calcium sulfate levels in an EDR system, as compared to no chemical addition, are 150 percent if a sequestering agent is added and 275 percent if an antiscalant is added.

High Recovery Rate EDR

The costs associated with concentrate disposal may be the biggest roadblock to widespread inland desalination. Any attempt to reduce the volume of concentrate would result in reducing the cost of desalination, and in some cases the goal is zero liquid discharge (ZLD). While there are some newly developed nonthermal technologies for brine concentration (for example, using electrodialysis or forward osmosis) that have minimized the volume of a concentrate stream, the application of these technologies has not been demonstrated in real world cases or the process is very costly. Depending on the water chemistry of the brackish/saline feedwater, the concentrate stream is usually supersaturated with some salts. Reducing the volume of the concentrate/reject stream directly depends on the supersaturation level of the salts in that stream. Thus, the limiting factor that governs the recovery rate (i.e., ratio of product flow to feed flow) is the level of scale-prone salts.

In order to get to higher recovery rate in conventional systems, the designers should have a comprehensive understanding of the scale forming minerals. Common scale forming minerals are calcite (calcium carbonate), calcium sulfate, and halite (sodium chloride). However, in some special cases other salts such as barium sulfate, strontium sulfate, and calcium fluoride also could be a concern and may contribute in scale formation. Normally, in seawater the main limiting factor in the recovery rate is halite, while in brackish water cases, depending on the feed water chemistry, the limiting factor could be calcite (calcium carbonate) or calcium sulfate. Calcium carbonate and calcium sulfate are the dominant salts in the Tularosa Basin and in most of the aquifers in the Southwest United States. Sulfate is the dominant anion in the water resources.

The highest reported supersaturation level for $CaSO_4$ was reported as $6.9 \times K_{sp}$ for >300 hours of operation using sodium hexametaphosphate antiscalant with acid addition, employing Ionics (SUEZ) Mark III-4 spacers. While the maximum super-saturation level is well established for traditional antiscalants, there are newer antiscalant products in the market for which the level of super-saturation for specific salts has not been established for various desalination technologies. Recent research that has been performed on full-scale EDR unit at BGNDRF (Chastain 2011) demonstrates promising results using newly developed antiscalant/antifoulant products with higher calcium sulfate super-saturation levels. Increasing water recovery rate using newly developed antiscalant/antifoulant products has a high potential for reducing the cost of desalination by minimizing the volume of concentrate and minimizing costs associated with concentrate disposal.

Pressure Drop

The pressure drop through a membrane stack is the sum of the pressure drops through each hydraulic stage. It is dependent on spacer type, flow rate per stage, and number of cell pairs in each stage. Spacers from various manufacturers have different pressure–velocity profiles. For example, systems needing higher removal rates benefit from using spacers that increase velocity, which increases ion contact with membranes. Because both ED and EDR are low-pressure systems, usually 50 psi (345 kPa) maximum at the inlet, and pressure is lost at each stage, providing for pressure through the entire system is a major design constraint. Changes in the numbers and types of spacers as well as adjustments to feedwater flow are methods of managing pressure drop through systems.

Differential Pressure

Differential pressure is the difference in hydraulic pressure between the demineralized stream and the concentrate stream. The pressure of the demineralized stream is maintained 0.5 to 1 psi (3.4 to 6.9 kPa) higher than the concentrate stream to prevent contamination of the product water stream. Thus, if a cross-leak does occur, product water will move into the concentrate stream and not vice versa.

Water Transfer

Although the membranes are virtually impermeable to water, a small amount of water is electrically transferred through the membranes along with the ions. The amount of water transferred varies with membrane type, type of ion, and the number of ions transferred. The higher the salt concentration compared with that of the product water, the greater the water transfer.

Temperature Limits

Typical membrane stacks have an upper operating limit for water temperatures of 113°F (45°C). This limit is determined by the low-density polyethylene spacer material, which becomes less rigid at elevated temperatures. The use of higher-quality spacers can raise the maximum operational temperature to 140°F (60°C).

WATER RECOVERY

Water recovery is defined as the total amount of water produced from the total amount of water used. EDR systems offer design options that optimize recovery, as discussed in this section. Off-specification product recycle, phased reversal, increased reversal cycle times, or any combination of these functions can be incorporated in EDR systems to take advantage of a substantial increase in recovery and production rates at minimal costs. Chemical dosing can further increase recovery rates and lower waste disposal costs.

Concentrate Recycle

Recirculating the concentrate stream is called concentrate recycle. Flow rates of concentrate and demineralized water through the membrane stacks are essentially equal. Without recycling, this equal flow would lead to a recovery of demineralized water equal to half of the saline water treated. However, in most ED and EDR systems, the concentrated effluent is recycled to conserve water. When the recycle stream is eventually sent to waste, the concentration level of effluent is controlled by adding an equal volume of source water according to concentrate disposal regulations.

Off-Specification Product Recycle

When the polarity reverses in an EDR system, the concentrate compartments become demineralizing compartments, and vice versa. The salinity in what used to be a concentrate compartment changes from a high TDS level to the product TDS level. The time needed for this change to take place is equivalent to the time required for source water entering this compartment to pass through the entire membrane system, normally 30 s per hydraulic stage. Product water leaving the system at this time is normally diverted to waste because it does not meet specifications. However, if reversal of the outlet valves is delayed, the average salinity of the product water can still be less than the salinity of source water. This product water can then be recycled back to the system feed, eliminating the need to send it to waste. This process is known as off-specification product recycle (OSPR).

Phased Reversal

Phased reversal (PR) in EDR is based on having independent control of (1) reversal of the inlet valves, (2) the electrical polarity of each stage (electrodes), and (3) reversal of the outlet valves. It works as follows: Suppose there are three stages in a series. If the polarity was reversed on all three stages at once, the water contained in those stages at that instant would not meet product specifications when it flowed out of the final stage. In other words, three stages' worth of water would not meet product specifications. However, if the polarity was reversed on the second stage a certain amount of time *after* reversal on the first stage (to allow all the off-specification water to finish passing through the first stage), only one stage's worth of water would not meet specifications. The timing of the third stage's polarity reversal would be handled similarly. Thus, the total reversal water is contained in a single stage in the system instead of through the entire system. PR increases recovery by increasing the amount of demineralized water produced while keeping the total amount of water consumed constant. The volume of waste during PR varies from 2 to 4 percent of the product water volume.

Reversal Frequency

In some cases, membrane scaling in an EDR system can be effectively controlled even with a reduced polarity reversal frequency. Standard systems use 15 min/polarity cycle, but some systems can operate with a 30-min cycle. Cutting the frequency of reversal in half reduces the portion of off-specification water by 50 percent. This increased net production augments recovery while reducing power consumption.

Electrode Stream Recycle

When cathode and anode streams are combined, they chemically neutralize one another and have a pH and salinity close to source water. Gases are collected and vented off, and liquid is normally sent to waste. Because this electrode stream quality is similar to that of source water, it may be recycled through the feed tank to increase system recovery and reduce total wastewater.

Chemical Addition

In EDR systems that do not use chemical dosing to prevent scaling, the maximum allowable concentrate salinity is determined by the limiting solubility salt concentrations (e.g., calcium carbonate or calcium sulfate). If higher recoveries are desired, then chemical dosing is needed to combat scaling.

As necessary, an antiscalant can be added to control calcium sulfate scaling, and acid can be used to control calcium carbonate scaling. Only a small quantity of antiscalant (1–5 mg/L) is required because the streams are recycled. The saturation of calcium sulfate can then be increased from 150 percent of its value in the absence of chemical addition to 275 percent. Acid is required to maintain the LSI at +1.8 in the concentrate stream for calcium carbonate control.

OPERATING PRINCIPLES FOR DESIGN

ED is a process in which ions are transferred through semipermeable ion exchange membranes from a less concentrated to a more concentrated solution by means of a direct electric current. There are four primary operating principles controlling ED, including

- dissociation of salts in water
- membrane properties
- Faraday's law
- Ohm's law

More details are provided in several texts (Bard & Faulkner 2001; Benjamin & Lawler 2013; Davis et al. 1997; Green & Perry 2007; Metcalf & Eddy 2014; Newman & Thomas-Alyea 2004; Strathmann 2004; Tanaka 2015).

Dissociation

Ions are formed by the dissociation of salts, minerals, alkalies, and acids in water. For example, sodium chloride dissolves and dissociates in water to form sodium ions (Na^+) and chloride ions (Cl^-). These ions conduct electricity through the solution. The conductivity is dependent on the concentration of ions, the ionic composition, and the temperature of the solution.

Membrane Properties

The most important properties of an ion exchange membrane are ion selectivity (that is, selection of specific ions for removal) and electrical conductivity. Ion selectivity of cation and anion membranes is greater than 90 percent for solutions with a normality of less than 0.5*N* up to 0.7*N*, but ion selectivity decreases at higher concentrations. Ion exchange membranes that have a low electrical resistance consume less energy, which increases system efficiency. As the concentration of the solution increases, resistance decreases.

The organic content of source water is an important factor in membrane selection. Organic acids, which may be present in brackish water sources, decrease membrane efficiency. As they adsorb to the anion exchange membrane, they increase membrane resistance and lower ion transfer. The subsequent need for cleaning can be decreased with the selection of appropriate membranes (Hamada 1993).

Faraday's Law

Faraday's law is the basis for determining the amount of electrical current needed in an ED process to transfer a specific quantity of salt. It states that the passage of 96,500 A of electric current for 1 s will transfer 1 g equivalent of salt. The quantity 96,500 A·s is called a *faraday*. In a sodium chloride solution, 1 faraday will transfer 1 g equivalent of sodium ions to the cathode and 1 g equivalent of chloride ions to the anode at 100 percent current efficiency. In other words, in the case of NaCl, the atomic weight of sodium is 23, which is 1 g equivalent. The atomic weight of chloride is 35. One faraday will transfer 23 g of sodium to the cathode and 35 g of chloride to the anode.

The following equation forms the basis for calculating the electric current needed for transferring ions. When used in ED calculations, Faraday's law is expressed for each cell pair as

$$I = \frac{F^* \times Fd \times \Delta N}{e \times N^*}$$

Where:

I = direct electric current, in A
F^* = Faraday's constant = 96,500 A·s/equivalent/cell pair
Fd = flow rate of the demineralized stream through the membrane stack, in L/s
ΔN = change in normality of demineralized stream between inlet and outlet of the membrane stack
e = current efficiency
N^* = number of cell pairs

Ohm's Law

Ohm's law states that the potential, E, of an electrical system is equal to the product of current I and the system resistance R. In other words,

$$E = I \times R$$

where E is expressed in volts, I is expressed in amperes, and R is expressed in ohms.

Voltage requirements for an ED or EDR system depend on current and resistance. Current is determined by the use of Faraday's law, whereas resistance is determined by the components of the stack and by source water quality. Resistance is affected by temperature and conductivity, and is measured with an alternating current (AC) resistance bridge. Resistance in the system is calculated from feedwater conductivity, with corrections made for the surface area of the membranes. For example, a highly conductive water will need more surface area contact with the membrane to remove ions. Each type of membrane and each plant has unique characteristics that affect resistance. The components that contribute to membrane resistance are expressed in the following equation:

$$R_{cp} = R_{cm} + R_{am} + R_c + R_d$$

Where:

R_{cp} = resistance per unit area of one cell pair, in Ω/cm²
R_{cm} = resistance per unit area of cation membrane, in Ω/cm²
R_{am} = resistance per unit area of anion membrane, in Ω/cm²
R_c = resistance per unit area of concentrate stream, in Ω/cm²
R_d = resistance per unit area of demineralized stream, in Ω/cm²

Resistance is also affected by other factors including temperature, ionic species, and ionic concentration. High temperatures and solution concentrations lower the solution resistance as well as the stack resistance. For example, a 1°F (0.55°C) change in temperature can result in about a 1 percent change in stack resistance.

Current Efficiency

Current efficiency refers to the percentage of total current that is effectively transferring ions. In ED and EDR applications, the current efficiency e for the transfer of salts can be calculated based on the expressions in Faraday's law as follows:

$$e = \frac{F^* \times F_d \times \Delta N \times 100}{I \times N^*}$$

As an example, consider the following sample operating readings:

Source water (normality)	0.0480*N*
Product water (normality)	0.0078*N*
Demineralized flow rate (*Fd*)	9,765 L/h (43.0 gpm)
Current (*I*)	25.5 A
Number of cell pairs (*N**)	450
Faraday's constant (*F**)	26.8 A·h (96,500 A·s)

The difference between feed normality and product normality is

$$\Delta N = (0.0480 - 0.0078) = 0.0402$$

Therefore, the current efficiency in this case is

$$e = \frac{26.8 \times 9{,}765 \times 0.0402 \times 100}{25.5 \times 450} = 91.7 \text{ percent}$$

Commercial membrane stacks normally have a minimum current efficiency of 90 percent, which means that 90 percent of the total current applied to the membrane stack is effectively transferring ions from the demineralized stream to the concentrate stream. The remaining 10 percent of the current can be attributed to loss resulting from cross leakage, current leakage, and back diffusion.

REFERENCES

Bard, A.J. and Faulkner, L.R. 2001. *Electrochemical methods: Fundamentals and applications*. Wiley, New York.

Benjamin, M.M. and Lawler, D.F. 2013. *Water Quality Engineering*. Wiley, Hoboken, N.J.

Chastain, M. 2011. New Research Changes Beliefs About EDR. https://iee.nmsu.edu/new-research-changes-beliefs-about-edr/.

Davis, T.A.; Genders, D.; Pletcher, D. 1997. *A First Course in Ion Permeable Membranes*. Electrochemical Consultancy, Hants, England.

Green, D. and Perry, R. 2007 (8th ed.). *Perry's Chemical Engineer's Handbook*. McGraw-Hill.

Hamada, M. 1993. Brackish Water Desalination by Electrodialysis. *International Desalination and Water Reuse Quarterly*, 2(4):8–15.

Metcalf & Eddy | AECOM. 2014. *Wastewater Engineering: Treatment and Resource Recovery*. McGraw Hill.

Newman, J.S. and Thomas-Alyea, K.E. 2004. *Electrochemical systems*. Wiley, Hoboken, N.J.

Pontius, Nancy. 1993. *Back to Basics Guide to Corrosion Control for Lead and Copper*. Denver, Colo.: AWWA.

Siwak, L.R. 1993. Here's How Electrodialysis Reverses and Why EDR Works. *International Desalination and Water Reuse Quarterly*, 2(4):16.

Strathmann, H. 2004. *Ion-exchange membrane separation processes*. Membrane science and technology series, Elsevier, Boston.

Tanaka, Y. 2015 (2nd ed.). *Ion Exchange Membranes: Fundamentals and Applications*. Elsevier, Amsterdam, Netherlands.

This page intentionally blank.

Chapter 4

Facility Design and Construction

By Val S. Frenkel and Patrick Girvin

The electrodialysis (ED) and electrodialysis reversal (EDR) systems are designed to remove dissolved minerals from water by the migration of ions through a stack of anion and cation exchange membranes under the force of a direct current (DC) electric field.

Each membrane stack typically consists of the following components:

- anion and cation exchange membranes
- intermembrane and electrode flow spacers
- electrodes
- plastic end blocks
- metal end plates
- tie rods and protective covers

The exact number of ion exchange membranes and spacers is determined by specific application

Custom-engineered and standard prepackaged plants are available from various ED/EDR manufacturers. Hydraulic capacity, percent recovery, level of salt removal, and membrane selection determine exact equipment specifications. Figures 4-1 and 4-2 illustrate the wide variety of packaged plants. The plant in the first figure has a processing capacity of 120 to 150 gpm (650 to 820 m^3/d); the second plant can handle up to capacity 13.6 mgd (3,600 m^3/d).

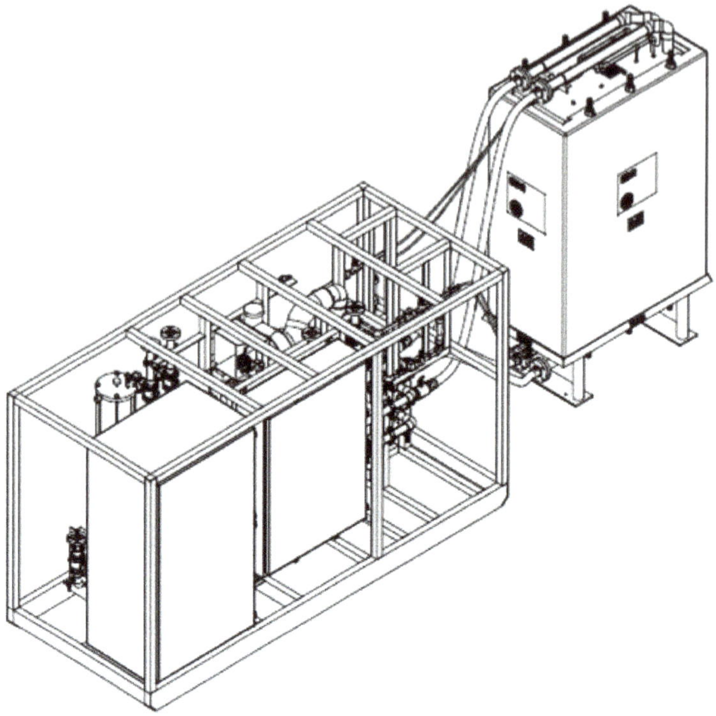

Source: SUEZ Water Technologies & Solutions

Figure 4-1 Plant with a processing capacity of 120–150 gpm

Source: SUEZ Water Technologies & Solutions

Figure 4-2 Plant in Suffolk, Va., that treats up to 1.3 mgd (5,000 m3/d). Total plant capacity 13.6 mgd

Hydraulic Controls

The major components of an ED or EDR hydraulic system typically include

- low-pressure centrifugal feed pump, stainless-steel
- low-pressure concentrate recirculation pump, stainless-steel
- cartridge filter housing, fiberglass reinforced plastic (FRP) or stainless steel (SS)
- corrosion-resistant piping, often PVC
- flow-control valves
- pressure reducing valve (PRV) for gradient pressure control between dilute and concentrate streams
- electrically or pneumatically actuated reversal valves (EDR only)

Table 4-1 lists one manufacturer's recommendations for EDR unit sizes and electrical connection requirements based on the desired processing capacity. Systems are available in a variety of sizes, and recommendations may vary among manufacturers.

Pumping pressures of 50 to 75 psi (345 to 515 kPa) provide adequate pressure for most systems, once head loss from varying system designs and sizes is taken into account. The low operating pressures for ED and EDR systems make it possible to use standard plastic piping and fittings as well as other common water treatment materials, reducing construction costs as well as corrosion concerns.

Instrumentation

The following instrumentation is required as a minimum to properly operate ED and EDR systems:

- conductivity analyzer in feed and product water
- pressure transmitters on pump discharge, stack inlet lines
- differential pressure transmitters on stack inlet and outlet
- flowmeters on product line and concentrate blow down/waste line
- temperature on feed line
- voltage and amperage monitoring on each individual ED/EDR stack

Table 4-1 Example EDR system sizes and power requirements

Capacity, gpm (m^3/h)	Total Stacks / # Stages	Footprint Size, inches (meters)	Connection Required, kVA
165–325 (37.5–73.8)	4 / 2	90 × 309 (2.3 × 7.9)	140
165–270 (37.5–61.3)	6 / 3	90 × 375 (2.3 × 9.5)	107
325–655 (73.8–148.8)	8 / 2	169 × 493 (4.3 × 12.5)	276
325–545 (73.8–123.8)	12 / 3	169 × 625 (4.3 × 15.9)	209
485–985 (110.2–223.7)	12 / 2	270 × 493 (6.0 × 12.5)	380
485–820 (110.2–186.2)	18 / 3	270 × 625 (6.0 × 15.9)	285
645–1315 (146.5–298.7)	16 / 2	270 × 493 (6.0 × 12.5)	542
645–1090 (146.5–247.6)	24 / 3	270 × 625 (6.0 × 15.9)	397

Performance, number of stages and cell pairs, recovery, and power consumption are dependent on inlet feed water quality and temperature. Typical power consumption is 2-4 kW·h/1,000 gallons of product water.

Source: SUEZ Water Technologies & Solutions

The following instrumentation is not mandatory while recommended to allow monitor and control certain parameters in specific applications:

- pH monitoring on brine stream to validate acid addition
- conductivity analyzer in concentrate water
- ORP Analyzer to monitor residual chlorine in feed water
- turbidity analyzer to monitor particulate matter load in the feed water

Electrical Controls/Safeguards

The major components of an ED or EDR electrical subsystem include

- adjustable DC power supply
- rectifiers to convert alternating current (AC) power to DC power
- internal control system with controls, reversal timing (EDR only), and alarms

Table 4-2 provides a sample engineering summary, including power consumption information, for the EDR system used in Buckeye, Ariz. (Carpenter and Geishecker 1989). Because no surface water is available, wells provide domestic water for more than 4,000 residents (1988). Well water salinity varies from 1,500 to 4,000 mg/L total dissolved solids (TDS), more than three times the EPA's secondary standard for TDS.

Table 4-2 Engineering summary for EDR systems used in Buckeye, AZ, and Magna, UT

	Buckeye A	Buckeye B	Buckeye System	Magna System
Number of plants	1	1	2	4
Number of lines per plant	2	3	5	8
Number of stages per line	3	3	3	2
Source water TDS—*mg/L*	1,587	1,587	1,587	1,555
Product water TDS—*mg/L*	246	328	287	500
Concentrate TDS—*mg/L*	6,951	6,623	6,787	7,500
Percent salt removal	84.5	79.3	81.9	67.8
Product water capacity—*gpd*	300,000	600,000	900,000	6,000,000
Source water capacity—*gpd*	375,000	750,000	1,125,000	7,059,000
Concentrate capacity—*gpd*	75,000	150,000	225,000	763,000
Percent water recovery	80.0	80.0	80.0	85.0
Source water temperature—*°F*	80	80	80	60
Concentrate Langelier Saturation index	+1.79	+1.73	+1.76	+2.54
Concentrate percent calcium fluoride (CaF_2) saturation	214.3	218.4	216.4	332
Total electrical consumption, *kW·h/1,000 gal*	4.0	3.9	3.95	3.1
DC power consumption—*kW·h/1,000 gal*	1.8	1.9	1.85	1.7
Pump power consumption—*kW·h/1,000 gal*	2.2	2.0	2.1	1.4
Source water	Well water	Well water	Well water	Well water
Pretreatment	Not required	Not required	Not required	Not required
Product use	Drinking water	Drinking water	Drinking water	Drinking water

Source: SUEZ Water Technologies & Solutions

An ED system implemented in 1962 was the first municipal desalination plant in Arizona, and it continued to supply water for Buckeye for 20 years. As a result of population growth and age of equipment, however, the system was replaced in 1988 by two EDR units described in Table 4-2, which were sized to allow for a 50 percent population growth. Raw water with an average TDS of 1,587 mg/L flowed through two EDR plants, one treating 300,000 gpd to 246 mg/L average TDS, the other treating 600,000 gpd to an average of 328 mg/L TDS. The product waters from the two systems (A and B) were then blended to produce 900,000 gpd of 287 mg/L average TDS. Plant TDS removals averaged 81.9 percent, and electrical consumption averaged 4 kW·h per 1,000 gal (3,785 L) of product water for Buckeye's new water treatment system.

In the early 2000s, Magna, Utah, realized that they had both arsenic and perchlorate in their source water. After settling on the EDR process, they installed a 2-stage system to meet the regulations for arsenic and perchlorate levels in drinking water supply.

The system is sized for 6 mgd and has the option to expand to a 3- or 4-stage system as the perchlorate levels rise in the incoming wells as expected. The system went online in 2008.

Chemical Cleaning System

An ED or EDR unit is equipped with a clean-in-place (CIP) system to allow periodic flushing of the membrane stack and piping with an acid solution. A chemical feed pump and storage tank form the main components of the CIP system. The ED/EDR CIP system is similar to the RO CIP system and contains the following components:

- booster pump (smaller systems use the concentrate pump on skid)
- CIP tank with heater
- chemical feed pump or drum pump to add cleaning chemical to tank
- flow meter
- cartridge filter (optional)

Installation – Building Details

ED and EDR equipment must be installed in a shelter to protect it from weather and blowing dust. In cases where electronic components are exposed to extremely low or high temperatures, the shelter should have heating or cooling to reduce corrosion and increase the operating life of the system. The area size required for the installation will depend on capacity, manufacturer, peripheral equipment, and open space around equipment required for maintenance.

Floor requirements. The building floor, normally made of poured concrete in accordance with local building codes, should provide adequate drainage for water discharged from the system during normal operation. The floor should be sloped to drain, except for the leveled stack area. Oftentimes, stacks are elevated off the ground slightly to prevent water pooling and corrosion of the base plates. Concrete should be treated with a corrosion-resistant epoxy resin coating to withstand the effects of spills.

Waste lines. During normal operation, all wastewaters must be drained away from the unit. This includes concentrate, off-specification product, electrode waste, and neutralized CIP waste. Any incidental spillage or wash-up water should be collected by the floor drains. All discharge water pipes installed in the skid are standpipes. These standpipes prevent the siphoning of water from the ED or EDR unit in standby mode and regulate back pressure on the stack(s) for proper plant operation and product water quality and quantity.

Drainage. During normal operation and maintenance, water may accumulate in the stack area (due to stack wash downs). Any incidental spillage or wash-up water should be collected by the floor drains. Drains should be placed so that there is no accumulation of water in the stack area.

Pipe trenches and racks. Depending on the requirements of the installation, pipes carrying source water to the unit or product water and wastewater from the unit can be either laid in trenches under the building floor or supported in overhead racks. Specific installation requirements should be discussed with the manufacturer. All materials used in trenches and racks should be acid resistant since there might be direct contact to low pH CIP solution.

Equipment servicing. On all sides of the skid(s) and stack(s), 3 to 5 ft (1 to 1.5 m) of clear space should be allowed for service and maintenance access. An overhead clearance is also required above the stack for maintenance (electrode and membrane replacements). The overhead clearance should be able to accommodate the crane as well as the top endplate.

Chemical Storage and Handling

During normal operation of an ED or EDR plant, chemicals used in operation, cleaning, and maintenance include

- sodium hexametaphosphate or other antiscalants, for scaling prevention
- chlorine, for disinfection or for oxidation of contaminants in feed water
- acid solution, for cleaning and scale removal in the stacks and electrode compartments
- salts, for cleaning organic deposits from membranes
- caustic solution, for cleaning organic deposits from membranes

Chemicals should be placed in a chemical storage room according to MSDS sheet and storage directions, and all safety precautions should be observed when chemicals are being handled.

Chemical room design is very important and requires careful review. As hydrochloric acid (HCl) is a common component in EDR systems, the chemical room should be designed with the fuming of this chemical in mind. Additionally, acid and chlorine should never be stored near each other.

Equipment Installation

The hydraulic skid is the first item brought into the plant. Once it is properly placed, it should be leveled. It is necessary to bolt the skid to the floor only if required by local building or seismic codes. After the hydraulic skid is in place, membrane stacks are installed according to manufacturer specifications. Stack interconnections and piping are then installed.

Each ED or EDR unit has several piping connections to be installed once the skid is in place. Piping for source water to the unit and for product water and discharged water from the unit must be connected. Piping may be overhead or in pipe trenches, with all piping and valves made from materials resistant to corrosion.

Inlet piping should be fitted with a suitable valve and strainer. This valve allows the system to control the inlet water and shut the unit off in case of excessive pressure. Strainers are installed for startup to prevent any debris from the piping including PVC chips from entering the stack. The strainers can be removed after the start up testing is completed.

Electrical Connections

A power-supply cable sized for the rating of the ED or EDR unit is brought into the electrical enclosure and connected to the main circuit breaker. The AC power supply (110, 220, 380, 460, 480 V, and 50 or 60 Hz, depending on the system) should have a separate circuit breaker or isolator in a location away from the unit.

A grounding cable should also be connected to the rear of the electrical enclosure to protect equipment from stray voltages. Additionally, each of the ED and EDR stacks and skids should be grounded to one common ground.

ED and EDR units are normally equipped for automatic operation using a level control in a product storage tank. Control wiring is installed according to the manufacturer's instructions and must consist of a voltage-free contact. No other external control wiring is required.

ED Cost Factor – Source Water Quality

Source water quality and criteria to the treated water quality are the primary factors determining capital and operating costs and the level of product recovery for ED/EDR systems. The type and quantity of ionic constituents in the source water define the type of system, its size, its pre- and post-treatment requirements, and the appropriate concentrate disposal methods. Recovery is limited by the scaling salts in the water including calcium carbonate and calcium sulfate. Temperature of water is another aspect that must be considered in ED and EDR systems because temperature affects conductivity and the ions mobility in the water, promoting or reducing removal and product water quality as a result.

ED Cost Factor – Electricity

Electricity is the main operating cost of an ED/EDR system, and the amount of electricity required is directly proportional to the amount of salt to be removed, as shown below in Figure 4-3. This illustrates why ED tends to be more competitive on brackish feed waters than on seawater or high TDS brine wastes.

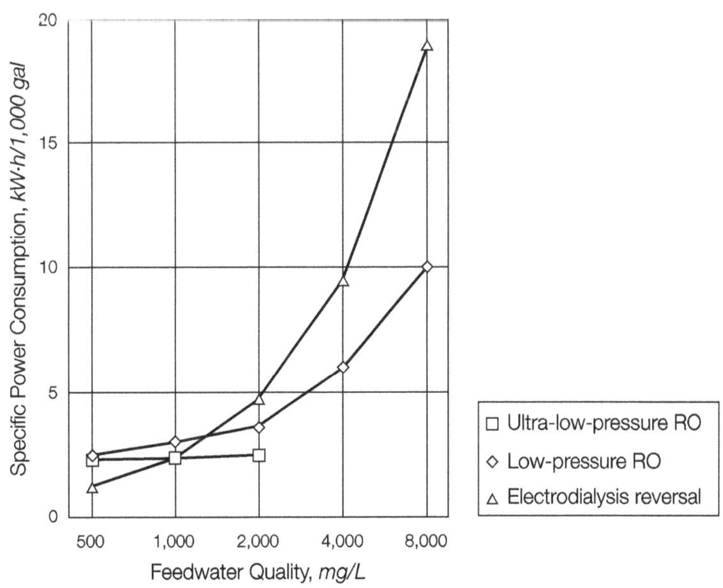

Figure 4-3 Effect of feed water quality on power consumption

ED Cost Factors – General

Some costs are generally applicable to various types of water treatment systems, such as costs involved in land and site development, wells or intake structures, storage, pumping, construction, and financing. Planning for anticipated system demands may make a larger system a sound investment due to the economy of scale because unit cost decreases as the system size is increased. The economy of scale, growth planning, and competition in the business environment all affect the sizing and costing of a water treatment plant.

Recovery Rates

Recovery rates for ED and EDR typically exceed 85 percent, depending on source water scaling salts, and they can be as high as 94 percent. Higher water recovery rates minimize system blowdown, pretreatment, and wastewater discharge volume in ED and EDR systems. This low wastewater volume may allow easier permitting and a smaller waste disposal system. However, smaller volume may concentrate wastewater contaminants, affecting permit requirements. Generally, low-pressure pumping, long membrane life, and low energy requirements decrease operations and maintenance costs in ED systems. EDR systems lower these costs further as a result of the "self-cleaning" effect of polarity reversal.

REFERENCE

Carpenter, F. and Geishecker, E. 1989. A Quarter Century of Electrodialysis Desalting: The Buckeye, Arizona, Experience. In Proc. 108th Annual Conference of AWWA held in Los Angeles, Calif., July 1989. Denver, Colo.: AWWA.

Chapter 5

Operation and Maintenance

By Jamie Vinsant and Rick Bond

Operation and maintenance (O&M) of ED and EDR equipment varies with each system manufacturer as systems are not standardized across industry and are manufacturer specific. The proper vendor operation manual should be consulted for the equipment being considered. This chapter provides general instructions to indicate the extent of O&M necessary for ED and EDR systems.

OPERATION PROCEDURES

Those implementing ED and EDR systems typically need to take into account the information indicated in Figure 5-1 when starting and operating a system. The concentrations of the feedwater constituents, the operating conditions, and recovery associated with a particular ED or EDR setup are vital considerations. It is important to note that design conditions may vary considerably depending on application and system supplier.

Most ED and EDR systems are designed with fully automated control systems incorporating data acquisition systems to continuously monitor the system performance.

Control settings and operating parameters should be checked at least once a day and logged on the data sheet recommended by the supplier. Because ED/EDR is an unsteady state process, the data should be recorded at approximately 75 percent of the time through the polarity to maintain consistency in the data. Trends of system operation that vary from normal for that system should be noted. Corrective action necessary to maintain the system within appropriate parameters should be taken immediately. A typical operation log sheet for an EDR unit is shown in Figure 5-2. This applies to a three-stage EDR system with multimedia filters to filter out contaminants down to approximately 20-micron nominal in size. These filters are followed by a cartridge filter ahead of the membrane stages. The rating of the cartridge filters is generally less than 10 microns but this will be specified by the system supplier. The voltage and amperage for each stack in the system should be monitored and recorded during each polarity.

Feedwater Analysis:

Sodium	_______ mg/L
Calcium	_______ mg/L
Magnesium	_______ mg/L
Chloride	_______ mg/L
Potassium	_______ mg/L
Bicarbonate	_______ mg/L
Sulfate	_______ mg/L
Nitrate	_______ mg/L
Total dissolved solids	_______ mg/L

Electrodialysis Design Operating Conditions:

Source water temperature	______ °F	______ °C
Source water salinity	______ mg/L	______ (μmhos)
Product water salinity	______ mg/L	______ (μmhos)
Normality	______ Source	______ Product
Production per 24-h day	______ gpd	
Total waste per 24-h day	______ gpd	
Concentrate makeup per 24-h day	______ gpd	
Electrical energy consumption per 1,000 gal product	______ kW•h/1,000 gal	
Dilute flow rate	______ gpm	
Concentrate makeup flow rate	______ gpm	
Stack inlet pressure	______ psi	
Stack outlet pressure	______ psi	
Electrical current and voltage		
Stage 1	______ A	______ V
Stage 2	______ A	______ V
Stage 3	______ A	______ V

Source: SUEZ Water Technologies and Solutions

Figure 5-1 Design concern checklist for EDR systems

Date: ______________ Time: ____________ Elapsed Hours: __________ By: __________

	Polarity +	Polarity —
Multimedia 1 In/Out (psi)	______/______	______/______
Multimedia 2 In/Out (psi)	______/______	______/______
Feed Pump Discharge (psi)	______/______	______/______
Cartridge Filter In/Out (psi)	______/______	______/______
Temperature In (°C)	______/______	______/______
Stack Pressure In/Out (psi)	______/______	______/______
Stack Differential In/Out (H_2O)	______/______	______/______
Electrode Pressure (psi)	____________	____________
Electrode Flow (gpm)	____________	____________
Dilute In (gpm)	____________	____________
Concentrate Makeup (gpm)	____________	____________
Volts Stage 1 A/B	______/______	______/______
Amps Stage 1 A/B	______/______	______/______
Volts Stage 2 A/B	______/______	______/______
Amps Stage 2 A/B	______/______	______/______
Volts Stage 3 A/B	______/______	______/______
Amps Stage 3 A/B	______/______	______/______
Feed Conductivity/pH	______/______	______/______
Product Conductivity/pH	______/______	______/______
Concentrate Conductivity/pH	______/______	______/______
Product Flow (gpm)	______/______	______/______
Product Flow Totalizer (gal)	______/______	______/______

Source: SUEZ Water Technologies & Solutions

Figure 5-2 Typical operation log for an EDR system

Accurate records and careful observation of the system performance parameters can provide advance warning on system operational challenges. The following sections provide examples of the potential challenges to be observed during ED/EDR operation and the possible causes to be considered.

Pressure

A steady increase in Inlet pressure, for example, may indicate fouling problems due to particulates or scaling caused by exceeding supersaturation of lower solubility compounds.

Amperage

At constant voltage, the amperage is dependent on stack resistance, which in turn depends on the following factors:

- feed water temperature
- feed water TDS/conductivity
- membrane resistance
- membrane spacer thickness
- age and condition of stack

As the temperature rises, the amperage will increase on the leading stack in a series possibly resulting in lower amperages in later stacks in the series due to changes in the individual stacks feed composition.

The stack resistance can be impacted by scaling within the stack, electrode failure, and organic fouling of the membranes (by natural organic matter or possibly by polymers used in upstream water treatment).

If a significant difference in amperage is experienced in one polarity compared to the opposite polarity (EDR systems), this could be due to a misassembled stack, scaling, or electrode failure.

Product Conductivity

The effect of temperature is opposite in an ED/EDR system as compared to an RO system. In an ED/EDR system, the product water TDS may increase with lowering temperature due to increased resistance, lower DC amperage, and slower transport of ions from the diluting (product) stream to the concentrating (reject) stream. In an RO system, the product water TDS decreases with lowering temperature, as slower diffusion means less salt will travel through the membrane with the water.

MAINTENANCE REQUIREMENTS

Routine preventive maintenance will ensure efficient operation of ED/EDR system as well as a long and reliable service. Suggested daily, weekly, monthly, and yearly maintenance steps for EDR systems are provided in this section, as well as guidelines for maintaining specific components of the system.

Daily Maintenance

1. Complete the system log sheet and verify that the operating settings are correct. Use the operation data sheet form (see Figure 5-2) and systems log sheet supplied by the vendor.

2. Check electrode stream flowmeters to verify that the electrodes are operating normally.
3. Schedule replacement of the pretreatment filter cartridges based on the inlet-to-outlet pressure differential, in accordance with the manufacturer's recommendations. Filters are typically changed at a pressure drop of about 15 psi (100 kPa).
4. Perform the electrode clean-in-place (CIP) procedure, as described later in this chapter, to maintain the electrode compartments clear of scale and fouling.

Weekly Maintenance

1. Probe the membrane stacks using the procedure outlined later in this chapter to detect problems early. This procedure will reduce component damage and replacement costs.
2. Inspect all unit piping, pressure instruments, sample tap fittings, and tubing for leaks. Repair as necessary.
3. Clean the outsides of the ED/EDR stacks of encrusted salt by washing them with clean water. DC power to the ED/EDR unit must be off while the stack sides are wetted down. Even with the DC power off, it is inadvisable to touch the side of the stack because electrical potential can be stored within the stack due to concentration differences. Once the stack has been effectively flushed, these stored electrical potentials will be dissipated.
4. Inspect the stacks in the area around the electrodes for signs of shorting or burning. The ED/EDR unit must be off when the stacks are being inspected or cleaned.
5. Check that all pressure, differential pressure, and flow-rate instruments are working properly. All pressures should read zero when the unit is off.
6. Check the oil level on all pumps with automatic oilers.
7. Measure the electrode rinse flows twice every week.

Monthly Maintenance

1. Clean, prime, and repaint any rusted areas of the system cabinets.
2. Check the spare parts inventory and order replacement parts that are short in stock.
3. Check the pump seals for any signs of leakage and repair them as necessary.
4. Check the degasifier blower for dirt on the housing and impeller wheel.
5. Perform the CIP procedure as outlined later in this chapter. The chemicals, frequency, and concentrations may be changed based on plant needs as operating experience is gained.

Annual Maintenance

1. Remove and inspect the grounding-rod assemblies. If rods are scaled, clean them by immersing them in 2 percent hydrochloric acid. Replace any damaged or corroded rods.
2. Drain the degasifier tank and flush out any accumulations of scale.
3. Lubricate pump motors as required.
4. Check the operation of the electrode CIP pumps.

5. Check the tightness of all electrical connections.
6. Calibrate all instruments and gauges.

Stack Probing

Stack probing requires working on stacks that are electrically live. The operator should *always* wear rubber gloves and rubber soled shoes. Never touch the side of the stack directly.

Probing the stacks should be done on a weekly to monthly basis, or as required for troubleshooting purposes. Stack probing can indicate scaling occurring within the stack or potentially an incorrectly assembled stack. A stack maintenance log sheet should be used when this procedure is performed. The operator needs the following equipment to probe the stacks: rubber gloves, safety glasses, a utility water hose, a hand-held voltmeter, and the stack probes.

Step 1 Stop the ED or EDR unit and wait for the flush cycle to complete.

Step 2 Remove the stack siding. Stack sides must be wet for a stack probe to be accurate. If the stack sides are encrusted with salt or are dry, wet the stack with water, paying special attention to the side being probed. The ED/EDR unit *must* be powered off while the sides of the stacks are being wetted down.

Step 3 Start the ED/EDR system in local control and wait until the product conductivity readings are in the designed range.

Step 4 With the stack probes spread slightly apart, press them firmly against the upper half side of the stack to check that a reading can be obtained.

Step 5 Switch the range of the voltmeter to "0–10 V DC." Place one lead on the top electrode plate and the other on the heavy cation membrane adjacent to the electrode spacer. This reading should be less than 10 V. A higher reading would indicate the electrode is scaled or worn.

Repeat for the bottom electrode.

Step 6 Place the probes 1 in. (25 mm) apart and take readings down the side of the stack between the electrodes, keeping a constant distance between the probes. In a normal stack, the meter readings will be approximately the same.

Step 7 On an EDR system, repeat the probe after the unit has reversed polarity and product readings are normal.

It is important to note that a stack probe is considered normal if no voltage probe results in a reading substantially higher than any other probe.

Electrode Clean-in-Place Process

In ED systems, acid is continuously fed into the electrode stream of the cathode to prevent scaling. In EDR systems, on the other hand, electrode clean-in-place (ECIP) is a routine preventive-maintenance procedure to remove scale or fouling from the electrode system. Each EDR unit has a dedicated ECIP pump used to inject a 1.5 percent (by weight) solution of hydrochloric acid from the day tank. Hydrochloric acid solution is injected into the electrode compartment automatically for 30 to 60 minutes (i.e., 2 to 4 polarity cycles) per 24 hours of operation.

Clean-in-Place Process

The CIP process is intended to remove scale or reversible fouling from the membrane stack and hydraulic piping. The chemical solution circulated through the stack depends on the type of contamination. This is a required procedure used in preventive maintenance and should be completed according to manufacturer's recommendations.

Only the chemical solutions approved by the membrane manufacturer should be used in the CIP process to avoid irreversible damage to the membrane. pH limits as provided by the membrane supplier must be strictly adhered to. The following chemical solutions are commonly used.

Hydrochloric acid solution. Periodic cleaning with a 2 to 5 percent hydrochloric acid (HCl) solution is the most frequently used method. The solution is used to remove carbonate scale as well as some metal deposits and light dirt. The procedure requires the use of the CIP tank, deionized water, and hydrochloric acid.

Alkaline sodium chloride solution. A 3 to 5 percent NaCl solution removes organic foulants, which are present in many surface waters, from the membranes. This procedure requires the CIP tank, deionized water, sodium chloride (food grade), sodium hydroxide (NaOH), and pH test paper (pH 8.0 to 10.0). This procedure is adapted from the brine/caustic cleaning procedure that is widely used for cleaning of organically fouled anion exchange resins. That procedure uses a much higher pH and is unsuitable for most anion exchange membranes.

For ED/EDR, the solution should be at least 3 percent NaCl (0.55*N* chloride) and have a pH approaching the maximum pH allowed by the membrane supplier. There are some anion membranes that would be damaged by the use of pH greater than 11. This solution should then be circulated through the system. After the alkaline NaCl application, the operator should flush the membranes with HCl to remove excess salt and any other precipitates that may have formed due to the high pH operation.

Chlorine solution. A 10- to 50-mg/L chlorine solution disinfects the membranes and hydraulic piping. This procedure requires the CIP tank, deionized water, and 5.25 percent commercial hypochlorite solution mixed at a rate of 0.25-gal (1 L) hypochlorite per 500-gal (1,890 L) water. Some membranes can be damaged by the application of chlorine, and the membrane supplier should be consulted for the recommended disinfection procedure.

Stack Repair and Disassembly

The membrane stack may be removed from the ED or EDR unit if stack repairs are necessary. Repairs can be completed manually with the assistance of an overhead crane. Membranes should be kept wet during the repair process. Blind flange connections should be attached to the remaining stacks to prevent them from drying. Specific manufacturer's instructions for stack removal should be followed.

When a problem arises, stacks can be disassembled in accordance with procedures recommended by the system's manufacturer. Generally, disassembly requires that each piece be removed separately, with the exception of the top electrode, which can be replaced without the removal of any of the membranes.

It is important to maintain correct component orientation and to store membranes in water. The stack should be rebuilt in the order it was disassembled; incorrect assembly can reduce performance or cause scaling.

Troubleshooting

Troubleshooting an ED or EDR system requires an understanding of the electrical, hydraulic, and membrane stack subsystems and their interrelationships. Isolating a problem to a specific subsystem makes troubleshooting easier. The manufacturer's guide can be used to locate and isolate a specific problem.

Readings from the log sheet should be checked against the design conditions to determine any anomalies. For example, pressure readings not at design levels may indicate a need for the pressure switches and transmitters to be calibrated.

Correct set points should be maintained on safety devices, such as pressure switches and conductivity controllers. Incorrect settings will defeat the purpose of the safety feature and make troubleshooting more difficult.

Alarm Conditions

Manufacturers may install detection systems capable of recognizing critical conditions that will shut down the unit immediately without flushing (i.e., without allowing the water already in the system to leave the system) and activate a horn. Typical critical alarms include high feed pressure or concentrate pump pressure. These conditions could indicate blockage or physical damage to the system. Because these are low-pressure systems, sudden high pressure could force water through the membranes and outside the membrane stacks. Under these circumstances, the normal flushing of the system with clean water to remove concentrate and prevent fouling, would not occur. Such flushing could increase the pressure within the system and cause greater damage than an immediate shutdown of the system.

When noncritical alarm conditions occur, the unit will shut down after flushing, the alarm horn will sound, and an alarm indicator will confirm the specific alarm condition. Noncritical alarm conditions will not cause damage to the membranes or components of the system. Typical noncritical alarms include the following:

- high product conductivity
- electrode pressure
- degasifier pressure
- rectifier high temperature
- suspected stack problem

SAFETY

Proper operation of an ED or EDR system requires an understanding of the safety issues involved in electrical operation, chemical handling, and mechanical systems.

Electrical Operation

Equipment shutdown. During normal operation an operator can shut down the ED or EDR unit by placing the system in manual and pressing the stop–start button. The unit will then enter a flush cycle. When this cycle is completed, the unit will be idle. In an emergency, however, the unit should be turned off at the main circuit breaker because the stop–start button will begin the flush cycle.

Grounding. All ED and EDR systems, including the stacks, should be connected to an electrical ground of earth potential to prevent electric shock from stray voltages. If the unit is skid mounted, a separate earth ground should be attached to the skid. Stacks should then be grounded by a wire to the skid or electrical cabinet. Grounding cables are provided by the manufacturer and should be installed so that they are not easily removed.

When equipment is moved, dismantled, or repaired, all electrical ground connections are should be restored.

High voltage. ED and EDR equipment use high voltage alternating current (AC) power. This AC voltage is converted to direct current (DC) by a rectifier for use in the operation of the membrane stacks. *DC voltage should never be applied to the stacks unless water is flowing through the stacks.*

The connecting bolts for the electrode tab should be tight and corrosion free. Overheating due to loose connections could damage the membrane stack.

DC amperage will rise and fall according to source water temperatures. To save power and prevent damage to the stack when source water temperatures rise, DC stack voltages must be lowered until DC amperage returns to normal settings. The DC stack voltage should not be increased as source water temperatures drop unless the manufacturer provides specific instructions to do so.

Membrane stack. The wet stack sides, electrode plates, or the electrode plate tab connections should not be touched when the unit is operating. Operators should always assume that the stacks are electrically alive. Protective clothing, such as rubber gloves, should always be worn during a stack probe. The equipment should not be washed down when the power is on.

Electrical panels. All control panels contain live voltage. Some systems are designed so that the main power must be off before the doors can be opened. Extreme care must be taken to avoid contact with electrically live components.

Only staff experienced in servicing the control panel should troubleshoot the panel. Equipment should never be operated with control doors open unless the cabinet is being repaired. This will protect personnel from electrical hazards and will also protect the electrical system from dust and water.

Chemical Safety

The degasifier tank separates the dissolved gases from the electrode waste streams. The system should be vented to the atmosphere at least 1 ft (0.3 m) above the highest peak of the building.

Personnel should always wear protective clothing (rubber gloves, apron, face shield, goggles, boots, and chemical-protection suit) when handling corrosive solutions or when working on the chemical pumps and supply lines. In addition, they should always wash down the work area after handling the chemicals and always clean up any chemical spills. When acid and water are mixed, the acid should *always* be added to the water. Adding water to acid can cause a violent explosion, spraying acid on the operator.

Mechanical Systems

The stack end plates should be removed when stack maintenance is performed. A hoist should be used because the plate is very heavy. It is important to always move the plate out of the way and set it down on a firm surface before proceeding.

A centrifugal pump should never be operated with either the discharge or suction valve closed. This can cause friction to the impeller and heat the trapped volume of water to a high temperature. If this heated water escapes, it can cause serious burns.

Rotating equipment should never be operated (specifically pumps, blowers, or fans) without their protective guards in place.

This page intentionally blank.

Appendix A

Abbreviations

A	amperes
AC	alternating current
A·h	ampere-hours
A·s	ampere-seconds
°C	degrees Celsius (centigrade)
CIP	clean-in-place
cm/s	centimeters per second
DC	direct current
ECIP	electrode clean-in-place
ED	electrodialysis
EDR	electrodialysis reversal
°F	degrees Fahrenheit
ft	feet
ft^2	square feet
ft/s	feet per second
g	grams
gal	gallons
gpd	gallons per day
gpm	gallons per minute
hp	horsepower
in.	inches
kPa	kilopascals
Ksp	solubility product

kW·h	kilowatt-hours
L	liters
L/h	liters per hour
LPRO	low-pressure reverse osmosis
L/s	liters per second
LSI	Langelier Saturation index
m	meters
m^2	square meters
m^3/d	cubic meters per day
MF	microfiltration
mgd	million gallons per day
mg/L	milligrams per liter
m	micrometers
mho/cm	micromhos per centimeter
S/cm	microsiemens per centimeter
min	minutes
ML/d	megaliters per day
mm	millimeters
N	normal
NF	nanofiltration
NPDES	National Pollutant Discharge Elimination System
ntu	nephelometric turbidity unit
Ω/cm^2	ohms per square centimeter
O&M	operation and maintenance
OSPR	off-specification product recycle
ppm	parts per million
PR	phased reversal
psi	pounds per square inch
psig	pounds per square inch gauge
PVC	polyvinyl chloride
RO	reverse osmosis
s	seconds
TDS	total dissolved solids
UF	ultrafiltration
ULP	ultra-low pressure
USEPA	US Environmental Protection Agency
V	volts

Additional Sources of Information

Allison, R.P. 1993a. Electrodialysis Membrane Performance Characteristics. In Proc. of the AWWA Membrane Processes Conference in Baltimore, Md. Denver, Colo.: AWWA.

Allison, R.P. 1993b. High Water Recovery with Electrodialysis Reversal. In Proc. of the AWWA Membrane Processes Conference in Baltimore, Md. Denver, Colo.: AWWA.

Allison, R.P. 1991. Surface and Wastewater Desalination by Electrodialysis Reversal. In Proc. of the AWWA Membrane Processes Conference in Orlando, Fla. Denver, Colo.: AWWA.

Bergman, R.A. and Lozier, J.C. 1993. Membrane Process Selection and the Use of Bench and Pilot Tests. In Proc. of the AWWA Membrane Processes Conference in Baltimore, Md. Denver, Colo.: AWWA.

Clark, S.W.; Parrotta, M.J.; Thompson, M.A.; Harrington, G.W. 1991. Membrane Technologies and Drinking Water Regulations. In Proc. of the AWWA Membrane Processes Conference in Orlando, Fla. Denver, Colo.: AWWA.

DuPont de Nemours, E.I. 1992. *Permasep Products Engineering Manual*. Wilmington, Del.: DuPont Company.

Ibrahim, E.A. and Taylor, J.S. 1993. Cost and Performance Evaluation of RO and EDR Pilot Plant Operation. In Proc. of the AWWA Membrane Processes Conference in Baltimore, Md. Denver, Colo.: AWWA.

Ionics Inc. 1993. *Bulletin No. 141*. Watertown, Mass.: Ionics Inc.

Lykins, B.W. Jr. 1991. Membrane Summary: Performance, Concerns and Regulations. In Proc. of the AWWA Membrane Processes Conference in Orlando, Fla. Denver, Colo.: AWWA.

Ma, J.Y.; Everest, W.R.; Erdman, D.A. 1993. EDR or RO — A Big Decision Facing a Small Southern California Water Utility. In Proc. of the AWWA Membrane Processes Conference in Baltimore, Md. Denver, Colo.: AWWA.

Meller, F.H. 1984. *Electrodialysis (ED) and Electrodialysis Reversal (EDR) Technology*. Watertown, Mass.: Ionics Inc.

Sharbat, A. 2016. *Development of Photovoltaic Electrodialysis Desalination System*. Desalination and Water Purification Research and Development Program Report No. 191, Bureau of Reclamation, U.S. Department of the Interior.

Strathmann, H. 2004. *Ion-Exchange Membrane Separation Processes*. Elsevier Science, Amsterdam, Netherlands.

Tanaka, Y. 2015 (2nd ed.). *Ion Exchange Membranes: Fundamentals and Applications*. Elsevier, Amsterdam, Netherlands.

Taylor, J.S.; Duranceau, S.J.; Barrett, W.M.; Goigel, J.F. 1989. *Assessment of Potable Water Membrane Applications and Research Needs.* Denver, Colo.: AWWA Research Foundation and AWWA.

Thompson, M.A. and Robinson, M.P. Jr. 1991. Suffolk Introduces EDR to Virginia. In Proc. of the AWWA Membrane Processes Conference in Orlando, Fla. Denver, Colo.: AWWA.

US Bureau of Reclamation. 1990. *Reverse Osmosis and Ultrafiltration Membrane Manual.*

Wolfe, T.D. 1993. Electrodialysis Design Approaches. In Proc. of the AWWA Membrane Processes Conference in Baltimore, Md. Denver Colo.: AWWA.

Glossary

Alternating current (AC): A periodic current for which the average value over a period is zero. AC is measured in amperes.

Amperage: The strength of an electric current, measured in amperes (coulombs per second); the amount of electric current flow, analogous to the flow of water in gallons per minute or liters per second.

Ampere (A): In Système International, that constant current that, if maintained in two straight parallel conductors of infinite length or negligible cross section and placed 1 meter apart in a vacuum, would produce a force equal to 2×10^{-7} newton per meter of length (adopted at the 9th General Conference on Weights and Measures in 1948).

Anion: A negatively charged atom or molecule that forms when an atom acquires one or more extra electrons. Anions may be present in solids and in solution in water or other solvents.

Anion membrane: An electrodialysis membrane that allows the passage of anions (but not cations) and is practically impermeable to water under typical electrolysis system working conditions. An anion membrane is also called an *anion transfer membrane* or an *anion exchange membrane.*

Bipolar membrane: An ion exchange membrane composed of a layer of anion exchange material in contact with a layer of cation exchange material. Applying a DC current to a bipolar membrane results in splitting of water inside the membrane and generation of hydrogen (H^+) and hydroxyl (OH^-) ions.

Cation: A positively charged ion (e.g., H^+ or Zn^{2+}) or radical (as NH_4^+) that migrates toward the cathode.

Cation membrane: An electrodialysis membrane that allows the passage of cations (but not anions) and is practically impermeable to water under typical electrodialysis system working pressure. A cation membrane is also called a *cation transfer membrane* or a *cation exchange membrane.*

Conductivity (specific conductance): A measure of the ability of a solution to conduct electrical current. Its value is inversely proportional to the solution's electrical resistance. The conductivity value is commonly used in water-desalting processes to evaluate desalting efficiency and to estimate the total dissolved solids concentration; the conductivity value of a water sample is multiplied by an empirical factor representative of the typical total dissolved solids/conductivity ratio for the specific type of water. The units of conductivity are often reported as micromhos per centimeter at 25°Celsius, but this is not a System International unit; multiplying such a value by 10^{-4} converts to the value units of siemens per meter.

Current: The speed of electrical charge movement, or the "flow rate" of electricity, measured in amperes.

Current density: As used in desalting, amperage per unit area flowing through a stack; current flow through a stack per unit cross-sectional area.

Current efficiency: In desalting processes, the theoretical power required for a particular job outcome compared to (i.e., divided by) actual power consumed.

Direct current (DC): A unidirectional electrical current that does not change in value over time (or changes so little that the change can be neglected). It is measured in amperes.

ED: See *electrodialysis.*

EDR: See *electrodialysis reversal.*

Electrical conductivity (EC): The reciprocal of the electrical resistance in ohms, measured between opposite faces of a 1-centimeter cube of an aqueous solution at a specified temperature. Units, in System International, are siemens per meter or square amperes cubic seconds per kilogram per cubic meter.

Electrical stage: In an electrodialysis system, a grouping of membrane cell pairs between one anode and one cathode.

Electrical staging: The addition of anode and cathode electrode pairs to an electrodialysis membrane stack to optimize the direct current electrical current efficiency. A single membrane stack may have one or more electrical stages.

Electrodeionization (EDI): A demineralization process employing a specialized electrodialysis membrane process containing mixed bed ion-exchange resin that can produce high-purity water. Feedwater passes through ion-exchange resin located in dilute-flow stream channels between parallel anion -exchange membranes and cation-exchange membranes in a direct current (DC) field. Feedwater ions move toward the electrodes of opposite charge and are transported through the ion-exchange resin and ion selective membranes to concentrate-flow channels. The DC electrical field is operated to split water into hydrogen and hydroxide ions near the membrane and resin bead surfaces that continuously regenerate the resin. The ions removed from the feedwater are discharged from the system in the concentrate flow stream.

Electrodialysis (ED): A desalting process driven by an electrical potential difference between oppositely charged electrodes. Ions are transferred by direct electric current flow through cation and anion membranes, depending on ion charge, from a less concentrated solution to a more concentrated one, leaving a demineralized stream.

Electrodialysis reversal (EDR): An electrodialysis process in which the electrical polarity of the electrodes is reversed on a set time cycle, thereby reversing the direction of ions in the system for membrane scaling and fouling control. See also *electrodialysis*.

Electromotive force (emf): Basic electrical force, measured in volts.

Electrotechnologies: Electricity-based water treatment technologies, including ozonation, ultraviolet light irradiation, electrodialysis, electrodeionization, and others.

Faraday constant (F): The quantity of electricity that can deposit (or dissolve) 1 gram-equivalent weight of a substance during electrolysis. It is equivalent to 96,489 coulombs. Thus, 1 faraday of electricity will reduce 1 mole of silver ion (Ag^{+}) or 0.5 mole of copper ion (Cu^{2+}) to the metal form.

Faraday's law: A law of electrochemistry stating that the amount of any substance dissolved or deposited in electrolysis is proportional to the total electric charge passed.

Fouling potential: In membrane process treatment, the degree to which the feed stream contains substances that may foul a membrane system and cause a decline in performance. Potential foulants include suspended and colloidal solids, metal oxides, microbial matter, organic compounds, and other solutes.

Hydraulic staging: A setup for allowing multiple passes of water between electrodes used in an electrodialysis process, through a sequence of subsequent membranes in a pressure-driven membrane system, or through filters used in a filtration system to achieve further treatment.

Ion-exchange membrane: An ion-exchange resin cast in flat sheet form and used in electrodialysis and electrodialysis reversal systems. Such a membrane is essentially impermeable to water. It has a fixed charge that allows it to pass ions of the opposite charge (counter-ions) but exclude ions of the same charge (co-ions). It is sometimes called an *electrodialysis membrane*. See also *electrodialysis*.

Ion-selective membrane: In the electrodialysis process, a membrane that allows specific ions to pass through more readily than other ions, especially one that is developed to remove one or more targeted ions more efficiently by having relatively high membrane permselectivity, even between ions of similar charge. An example is an anion-permeable membrane that readily allows bicarbonate ($HCO_3^{\ -}$) and chloride (Cl^-) to pass but has relatively high resistance to sulfate (SO_4^{2-}).

Langelier saturation index: The most famous of the calcium carbonate ($CaCO^3$) saturation indexes, the formula for the Langelier saturation index is based on a comparison of the measured pH of a water (pHa) with the pH the water would have (pHs) if at saturation with $CaCO^3$ (calcite form) given the same calcium hardness and alkalinity for both pH cases. Many of the other indexes found in the water treatment and corrosion literature are merely approximations. Several versions of the LSI exist in the literature. The version presented here is the one developed by J.R. Rossum and D.T. Merrill for the quadratic equation form of the pHs equation solution. This material has been updated using newer equilibrium constant and activity coefficient data. (Note that a simplified approximation equation was originally formulated by T.E. Larson and A.M. Buswell.) The basic formula for the Langelier saturation index is

$$\text{LSI} = \text{pH}_a - \text{pH}_s$$

It is interpreted in the following way: If LSI > 0, water is potentially scale forming (supersaturated with respect to $CaCO^3$). If LSI = 0, water is in equilibrium with $CaCO^3$, not tending to dissolve or precipitate it. If LSI < 0, water will potentially dissolve existing $CaCO^3$ deposits (undersaturated with respect to $CaCO^3$). The quantity pHs is found from the computed hydrogen ion concentration at hypothetical saturation ($[H+]_s$) by

$$\text{pH}_s = -\log_{10}([H^+]_s) - \log_{10} f_m$$

Where:

f_m = the activity coefficient for monovalent ions

The parameter $[H^+]_s$ is computed by the following quadratic equation:

$$\left[H^+\right]_s = \frac{-B \pm \sqrt{B^2 - 4AC}}{2A}$$

Where:

$$A = 1 - \frac{\left[Ca^{2+}\right]K_2}{K_s}$$

$$B = K_2\left(2 - \frac{\left[Ca^{2+}\right]\text{Alk}}{K_2}\right)$$

$$C = \frac{K_w K_2}{K_s}\left[Ca^{2+}\right]$$

$[Ca^{2+}]$ = calcium concentration, in milligrams per liter as $CaCO_3$
Alk = alkalinity, in milligrams per liter as $CaCO_3$
K_2 = solubility constant
K_s = equilibrium constant
K_w = $[H^+][OH^-]$

Limiting current density: The current density at which dramatic increases in electrical resistance are observed such as in an electrodialysis system. Limiting current density is the point where liquid phase ionic diffusion rate limitations result in substantial depletion of ion concentrations at the membrane surface.

Membrane: A natural or synthetic material that separates solids and liquids. Membranes used for water treatment are commonly synthetic organic polymers. In the case of pressure-driven membranes, the polymers are permeable to water (solvent) but reject solutes and/or particulate matter (depending on membrane type); for electrodialysis membranes, the polymers are designed to allow ions to pass but not water. See also electrodialysis; electrodialysis reversal; microfiltration; nanofiltration; reverse osmosis; ultrafiltration.

Membrane area: The total active area of a permselective membrane.

Membrane selectivity: The ability of a membrane to allow passage of only cations, anions, or molecules of a certain size. See also *permselectivity*.

Membrane stack: In an electrodialysis–electrodialysis reversal system, the vertical arrangement of multiple membrane cell pairs between electrodes.

Nernst-Planck equation: This mathematical relationship is used to describe the motion of a charged chemical species in a fluid medium. It describes the flux of ions under the influence of both an ionic concentration gradient and an electric field.

ohm (Ω): In Système International, the unit of electrical resistance, equivalent to meters squared kilograms per second cubed per ampere squared. One ohm is the electrical resistance between two points of a conductor when a constant difference of potential of 1 volt, applied between these two points, produces in this conductor a current of 1 ampere, with this conductor not being the source of any electromotive force.

Ohm's law: A fundamental law of electricity that relates electromotive force to current and resistance. Ohm's law states that 1 volt of electromotive force will cause 1 ampere of current to flow through 1 ohm of resistance. It is written as

$$E = IR$$

Where:

E = electromotive force, in volts
I = current flow, in amperes
R = resistance, in ohms

Oxidation: (1) A process in which a molecule, atom, or ion loses electrons to an oxidant. The oxidized substance (which lost the electrons) increases in positive valence. Oxidation never occurs alone but always as part of an oxidation–reduction (redox) reaction. The reduced substance gains electrons and thereby decreases in positive valence. (2) In electrodialysis, a chemical reaction that occurs at an anode and results in the loss of electrons from the anodic material. (3) In ion exchange, a specific attack on the crosslinking of the copolymer of an ion-exchange resin by an oxidant (e.g., chlorine, hydrogen peroxide (H_2O_2), or ozone (O_3)) leading to degradation (loss of structure of resin beads) and shortening of the resin life.

Permeability: A measure of the relative ease with which water flows through a porous material. A sponge is very permeable; concrete is much less permeable. Permeability is sometimes called *perviousness*.

Permeability coefficient: A coefficient expressing the rate of flow of a fluid through a cross section of permeable material under a hydraulic gradient or pressure gradient. If the fluid is water, the permeability coefficient is the hydraulic conductivity. A standard coefficient of permeability used in the hydrologic work of the US Geological Survey, known as the Meinzer unit (not a Système International unit), is defined as the rate of flow of water at 60° Fahrenheit (15.6° Celsius), in gallons per day, through a 1-foot cross-section under a hydraulic gradient of 100 percent.

Permselective membrane: A membrane used for separation of constituents in a fluid based on differences in one or more constituent properties, such as diffusion rate, solubility, electrical charge, or size and shape. A permselective membrane is also known as a *semipermeable membrane*.

Permselectivity: The measure of the selectivity of an ion-exchange membrane toward the passage of counter-ions and exclusion of co-ions. Ideal permselectivity means that the permeability of the co-ion is negligible compared to the permeability of the counter-ion.

Polarity reversal: In an electrodialysis reversal membrane process, the periodic change in electrode polarity of the electrodes, whereby the demineralizing compartments become concentrate compartments and the concentrate compartments become demineralizing compartments.

Polarization: (1) The concentration or depletion of ions in the thin boundary layer adjacent to an ion-rejecting membrane. (2) In the electrodialysis process, the point at which the current density is great enough to dissociate water molecules, forming hydrogen ions (H^+) and hydroxide ions (OH^-).

Resistance: A measure of a material's ability to impede the flow of electricity. Resistance is measured in ohms. See also *Ohm's law*.

Resistivity: A measure of the ability of a solution or any other bulk material to retard electrical current. Resistivity is inversely proportional to conductance and is commonly used in highly purified water treatment systems to quantify the degree of process water demineralization. Resistivity is measured in ohm-meters.

Semipermeable membrane: See *permselective membrane*.

Spacer: For a membrane treatment unit, a material that is designed to allow controlled fluid flow and is used to separate membrane layers, such as a feed-concentrate or permeate spacer for spiral-wound membrane or a concentrating compartment or demineralizing compartment spacer for an electrodialysis membrane stack.

Stack: In demineralization, a basic electrodialysis operating unit consisting of membranes, separator, electrodes, and other appurtenances necessary to make a complete operating unit.

Tortuous path: (1) A water flow path through channels that are constricted and marked by repeated twists, bends, and winding turns. (2) In an electrodialysis system, a water flow path in which spacers, turbulence promoters, or cross-traps are used to produce turbulence in the flow stream.

Transference number: The fraction of the total electrical current carried in an electrolyte by a given ionic species. Also called *ion transport number*.

Volt (V): In Système International, the unit of electrical potential, potential difference, and electromotive force, equivalent to meters squared kilograms per second cubed per ampere. One volt is the difference of electric potential between two points of a conductor, carrying a constant current of 1 ampere, when the power dissipated between these points is equal to 1 watt.

Voltage: The electrical pressure available to cause a flow of current (amperage) when an electrical circuit is closed. Voltage is measured in volts.

Voltage-driven membrane: An electrodialysis membrane.

Water transport: For membrane processes, the passage of water through a membrane. Water transport is desirable in reverse osmosis and nanofiltration and undesirable in electrodialysis.

This page intentionally blank.

Index

Note: *f.* indicates figure; *t.* indicates table

T

W

This page intentionally blank.

AWWA Manuals

- M1, *Principles of Water Rates, Fees, and Charges,* #30001
- M2, *Instrumentation and Control,* #30002
- M3, *Safety Management for Water Utilities,* #30003
- M4, *Water Fluoridation Principles and Practices,* #30004
- M5, *Water Utility Management,* #30005
- M6, *Water Meters—Selection, Installation, Testing, and Maintenance,* #30006
- M7, *Problem Organisms in Water: Identification and Treatment,* #30007
- M9, *Concrete Pressure Pipe,* #30009
- M11, *Steel Pipe—A Guide for Design and Installation,* #30011
- M12, *Simplified Procedures for Water Examination,* #30012
- M14, *Backflow Prevention and Cross-Connection Control: Recommended Practices,* #30014
- M17, *Fire Hydrants: Installation, Field Testing, and Maintenance,* #30017
- M19, *Emergency Planning for Water and Wastewater Utilities,* #30019
- M20, *Water Chlorination/Chloramination Practices and Principles,* #30020
- M21, *Groundwater,* #30021
- M22, *Sizing Water Service Lines and Meters,* #30022
- M23, *PVC Pipe—Design and Installation,* #30023
- M24, *Planning for the Distribution of Reclaimed Water,* #30024
- M25, *Flexible-Membrane Covers and Linings for Potable-Water Reservoirs,* #30025
- M27, *External Corrosion Control for Infrastructure Sustainability,* #30027
- M28, *Rehabilitation of Water Mains,* #30028
- M29, *Water Utility Capital Financing,* #30029
- M30, *Precoat Filtration,* #30030
- M31, *Distribution System Requirements for Fire Protection,* #30031
- M32, *Computer Modeling of Water Distribution Systems,* #30032
- M33, *Flowmeters in Water Supply,* #30033
- M36, *Water Audits and Loss Control Programs,* #30036
- M37, *Operational Control of Coagulation and Filtration Processes,* #30037
- M38, *Electrodialysis and Electrodialysis Reversal,* #30038
- M41, *Ductile-Iron Pipe and Fittings,* #30041
- M42, *Steel Water-Storage Tanks,* #30042
- M44, *Distribution Valves: Selection, Installation, Field Testing, and Maintenance,* #30044
- M45, *Fiberglass Pipe Design,* #30045
- M46, *Reverse Osmosis and Nanofiltration,* #30046
- M47, *Capital Project Delivery,* #30047
- M48, *Waterborne Pathogens,* #30048
- M49, *Quarter-Turn Valves: Head Loss, Torque, and Cavitation Analysis,* #30049
- M50, *Water Resources Planning,* #30050
- M51, *Air Valves: Air-Release, Air/Vacuum and Combination,* #30051
- M52, *Water Conservation Programs—A Planning Manual,* #30052
- M53, *Microfiltration and Ultrafiltration Membranes for Drinking Water,* #30053
- M54, *Developing Rates for Small Systems,* #30054
- M55, *PE Pipe—Design and Installation,* #30055
- M56, *Nitrification Prevention and Control in Drinking Water,* #30056
- M57, *Algae: Source to Treatment,* #30057
- M58, *Internal Corrosion Control in Water Distribution Systems,* #30058
- M60, *Drought Preparedness and Response,* #30060
- M61, *Desalination of Seawater,* #30061
- M62, *Membrane Applications for Water Reuse,* #30062
- M63, *Aquifer Storage and Recovery,* #30063

M65, *On-Site Generation of Hypochlorite,* #30065

M66, *Cylinder and Vane Actuators and Controls—Design and Installation,* #30066

M68, *Water Quality in Distribution Systems,* #30068

M69, *Inland Desalination and Concentrate Management,* #30069

M77, *Condition Assessment of Water Mains,* #30077

www.ingramcontent.com/pod-product-compliance
Ingram Content Group UK Ltd.
Pitfield, Milton Keynes, MK11 3LW, UK
UKHW052013041026
15590UKWH00008B/2316

9 781647 17069